Pre-Algebra Made Simple

Written by Wendy Freebersyser

Illustrated by Don O'Connor

Cover Design by Signature Design Group

Notice! Copies of student pages may be reproduced by the teacher for home or classroom use only, not for commercial resale. No part of this publication may be reproduced for storage in a retrieval system, or transmitted in any form or by any means—electronic, mechanical, recording, etc.—without the prior written permission of the publisher. Reproduction of these materials for an entire school or school system is strictly prohibited.

FS122009 Pre-Algebra Made Simple
All rights reserved—Printed in the U.S.A.
Copyright © 1999 Frank Schaffer Publications, Inc.
23740 Hawthorne Blvd.
Torrance, CA 90505

Introduction

Math is a part of our daily lives, and it is important for children to understand "that being able to explain and justify their thinking is important and that how a problem is solved is as important as its answer." (NCTM, 1989)

What is a variable?

How can I look at this problem another way?

What does it mean to solve an equation?

How is perimeter different from area?

Answering these and other questions are students' attempts to make sense of mathematic concepts.

Pre-Algebra Made Simple is the perfect opportunity to make pre-algebra fun, relevant, and interesting for any student. It has been written to provide activities that will promote the application of mathematics and algebra in the workplace and in other real-life situations. In order to master a given concept or procedure, students must understand how the various representations are connected. Students must learn to view mathematics as an integrated whole rather than as a series of factoids and unrelated topics.

The objective of *Pre-Algebra Made Simple* is to help students form bridges that link arithmetic and algebra and geometry. Also, students must understand how topics in arithmetic, algebra, and geometry are related to data analysis, probability, and statistics. To help achieve this goal, a wonderful variety of activities have been included and encompass a variety of different puzzles—some relating to real-life applications; others featuring riddlelike situations.

This book is divided into five sections. At the beginning of each section are teacher resource pages. These pages provide many related activities and problems that can be used to guide students through each section. One exciting aspect of these resource pages is the Chapter Group Projects. These projects incorporate technology and cooperative learning while helping students apply the concepts and skills presented in the section in order to solve real-life applications.

After each set of teacher resource pages are interesting and exciting student activity pages. Students can work these pages to practice their skills and gain a conceptual understanding of the topics in each particular section. The students will incorporate prior knowledge with real-world situations and integrate concepts across the curricular areas. Mathematical connections allow construction of bridges between the concrete and the abstract, linking conceptual and procedural knowledge. These bridges invite students to informally explore, conjecture, and develop mathematical generalizations.

The concepts covered in *Pre-Algebra Made Simple* are basic to most pre-algebra courses. Students will develop a conceptual understanding of pre-algebra topics and will practice skills relating to the following specific concepts: order of operations (numberness), solving equations, geometry, data analysis, probability, statistics, and pattern recognition.

Pre-Algebra Made Simple is an easy and fun way to develop students' interest in and understanding of valuable arithmetic and algebraic concepts. You will be excited to observe as your students discover how stimulating learning algebra can be!

Whole Numbers and Integers

Algebra is arithmetic with variables. Therefore, it is vital that students have a strong background in basic arithmetic skills. This section includes addition, subtraction, multiplication, and division; other number sense, including numeration and estimation; and the application of these operations and concepts in the workplace and other situations. Help students understand the concept of the unknown and solving for the value of the unknown to help them better understand variables. Allow students to complete several examples with your guidance. Be sure students gain a conceptual understanding of the concepts to the right before proceeding through the student activity pages (pages 3–27).

Present everyday situations to students in which they may use their new skills. For example, students can use their knowledge of whole numbers and integers as they balance a checkbook or figure out how much they make a week if they have a job. Help students observe the world in which they live and create their own connections that involve working with real numbers.

CONCEPTS

The ideas and activities presented in this section will help students explore the following concepts:
- adding real numbers
- subtracting real numbers
- multiplying real numbers
- dividing real numbers
- order of operations
- time
- money
- problem solving
- ratio and proportion
- percent

GETTING STARTED

- Have students determine how many prime numbers there are between 100 and 200. (Answer: 32) Students should list strategies, possible answers, and show the process that they used.

- Introduce the order of operations to the students. (**P**lease **E**xcuse **M**y **D**ear **A**unt **S**ally.)

    1. operations within **p**arentheses (grouping symbols)
    2. **e**xponents (powers)
    3. **m**ultiplication or **d**ivision from left to right
    4. **a**ddition or **s**ubtraction from left to right

    Put several problems on the board. Go over them as a class so students get an understanding of the order of operations.

QUICK MOTIVATORS

- Ask students to determine how many different ways there are to make change for a 50-cent piece.
- Play the Number Game—the game of ⁻5 involves using one or more operation symbols (+, −, x, ÷) along with three numbers to write an expression having a value of ⁻5. For example, for the numbers 15, 5, and 8, the expression would be 15 ÷ 5 − 8 = ⁻5. (Play the game with ⁻3, 8, 4; 9, ⁻1, ⁻4; 18, 3, 11.) (Answers: ⁻3 − 8 ÷ 4; 9 • ⁻1 − ⁻4; 18 ÷ 3 − 11)
- Ask students to explain the rules of addition, subtraction, multiplication, and division. Have them note similarities and differences.

FS122009 Pre-Algebra Made Simple • © Frank Schaffer Publications, Inc.

FUN WITH MATH

Put the problems below on the board to get students on track and into algebra.

- The multiples of 3 (to 30) are 3, 6, 9, 12, 15, 18, 21, 24, 27, 30. On the calendar below, circle the multiples of 4. (Answers: 4, 8, 12, 16, 20, 24, 28)

S	M	T	W	Th	F	S
		1	2	3	4	5
6	7	8	9	10	11	12
13	14	15	16	17	18	19
20	21	22	23	24	25	26
27	28	29	30			

- Put the secret code below on the board and challenge the students to crack it.

   1. Su + 6 = Sa
   2. Th + 4 = _____ (Monday)
   3. W + 8 = _____ (Thursday)
   4. Sa + 7 = _____ (Saturday)
   5. Sa – 7 = _____ (Saturday)
   6. M + 14 = _____ (Monday)
   7. F + 8 = _____ (Saturday)
   8. Tu + 5 = _____ (Sunday)
   9. Th – 2 = _____ (Tuesday)
   10. Tu + 21 = _____ (Tuesday)

- Find five sets of numbers to use in a game of ⁻5. Then work in groups to play the game of ⁻5. (See directions for ⁻5 under Quick Motivators, page 1.)

- Put the calendar below on the board along with the problems. Challenge the students to determine the answers and then write 5 of their own problems for a classmate to solve.

   1. 3 ↓ = 10
   2. 4 → = _____ (5)
   3. 6 ↘ = _____ (14)
   4. 11 ↗ = _____ (5)
   5. 8 ↑ = _____ (1)
   6. 15 ↖ = _____ (7)
   7. 27 ↖ ↑ → = _____ (13)
   8. 18 → ↑ ← ↖ = _____ (17)
   9. 22 ↑ ↑ ← ↘ = _____ (15)
   10. 4 ↖ ↓ → ↘ = _____ (26)

S	M	T	W	Th	F	S
						1
2	3	4	5	6	7	8
9	10	11	12	13	14	15
16	17	18	19	20	21	22
23	24	25	26	27	28	29

Chapter Group Projects

- Have students work in groups to design a school T-shirt. Each group is to get two price quotes from local companies. To determine the percentage of students that would purchase T-shirts, groups should take an informal survey. Each group should use this prediction to determine how many shirts to purchase to maximize profit. Have students present their results to the class and discuss differences and similarities in group findings.

- Give groups of students the problems below to solve using the Internet.
   1. Determine a historical event that occurred one billion seconds ago.
   2. Determine the time difference between St. Louis, Missouri, and Paris, France.
   3. Determine the time difference between Dallas, Texas, and Rome, Italy.

Mini Mottoes

The seven states listed below have one-word mottoes. The smallest state has the shortest motto of the seven! What are these mottoes?

To find out the mottoes, solve the problems below. Find the number that each letter represents. Fill in the blanks with the proper letters. (Hint: Some letters will not be used.)

K. ⁻481 + 800 = _____ **Y.** ⁻82 + ⁻63 + 91 = _____ **M.** 43 + ⁻200 + 36 + ⁻8 = _____

G. 26 + ⁻19 = _____ **V.** ⁻463 + ⁻841 + ⁻150 = _____ **W.** ⁻19 + 56 + ⁻27 + ⁻80 = _____

S. ⁻42 + ⁻56 = _____ **F.** ⁻864 + 853 + 29 = _____ **E.** 55 + ⁻44 + 20 + 13 = _____

D. ⁻98 + 69 = _____ **U.** 146 + ⁻146 + 384 = _____ **R.** 20 + 73 + 84 + 135 = _____

B. 127 + ⁻166 = _____ **A.** 44 + 38 + ⁻65 = _____ **X.** ⁻246 + 31 + ⁻57 + ⁻100 = _____

C. ⁻254 + 321 = _____ **P.** ⁻123 + ⁻56 + ⁻81 = _____ **Q.** 145 + ⁻83 + ⁻92 + ⁻61 = _____

I. 29 + ⁻13 + 46 = _____ **L.** ⁻5 + 43 + 30 + ⁻87 = _____ **O.** 87 + ⁻3 + 4 + ⁻18 + 27 = _____

N. 96 + 52 + ⁻83 = _____ **T.** ⁻28 + ⁻63 + ⁻81 + ⁻7 = _____ **H.** 56 + ⁻88 + 9 + ⁻361 + 59 = _____

California __ __ __ __ __ __
               44 384 312 44 319 17

Maine __ __ __ __ __ __
          ⁻29 62 312 62 7 97

New York __ __ __ __ __ __ __ __ __
             44 ⁻372 67 44 ⁻19 ⁻98 62 97 312

Rhode Island __ __ __ __
                ⁻325 97 ⁻260 44

Texas __ __ __ __ __ __ __ __ __
        18 312 62 44 65 ⁻29 ⁻98 ⁻325 62 ⁻260

Utah __ __ __ __ __ __ __ __
       62 65 ⁻29 384 ⁻98 ⁻179 312 ⁻54

Wisconsin __ __ __ __ __ __ __
          18 97 312 ⁻70 17 312 ⁻29

Addition of integers

The Name Game

Suppose that numbers are assigned to the letters of the alphabet as follows:

A = 1, B = 2, C = 3, D = 4, E = 5, and so on to Z = 26.

Using the code, find the numerical value of the names below by multiplying the values of its digits.

Example: ED = 5 • 4 = 20

1. Bob _____

2. Amy _____

3. Sue _____

4. Juan _____

5. Find the value of your name.
   _____
   Compare it with the value of the names of others in your class. Are there two names in your class that have the same value?

6. Find a boy's name that has three letters and a value of 140. (Hint: One letter is B.)

   _____

7. Find a girl's name that has three letters and value of 140. (Hint: One letter is the same as in #6.)

   _____

8. Find a girl's name that has four letters and a value of 700. (Hint: One letter is J.)

   _____

9. Find a boy's name that has four letters and a value of 1400. (Hint: One letter is Y.)

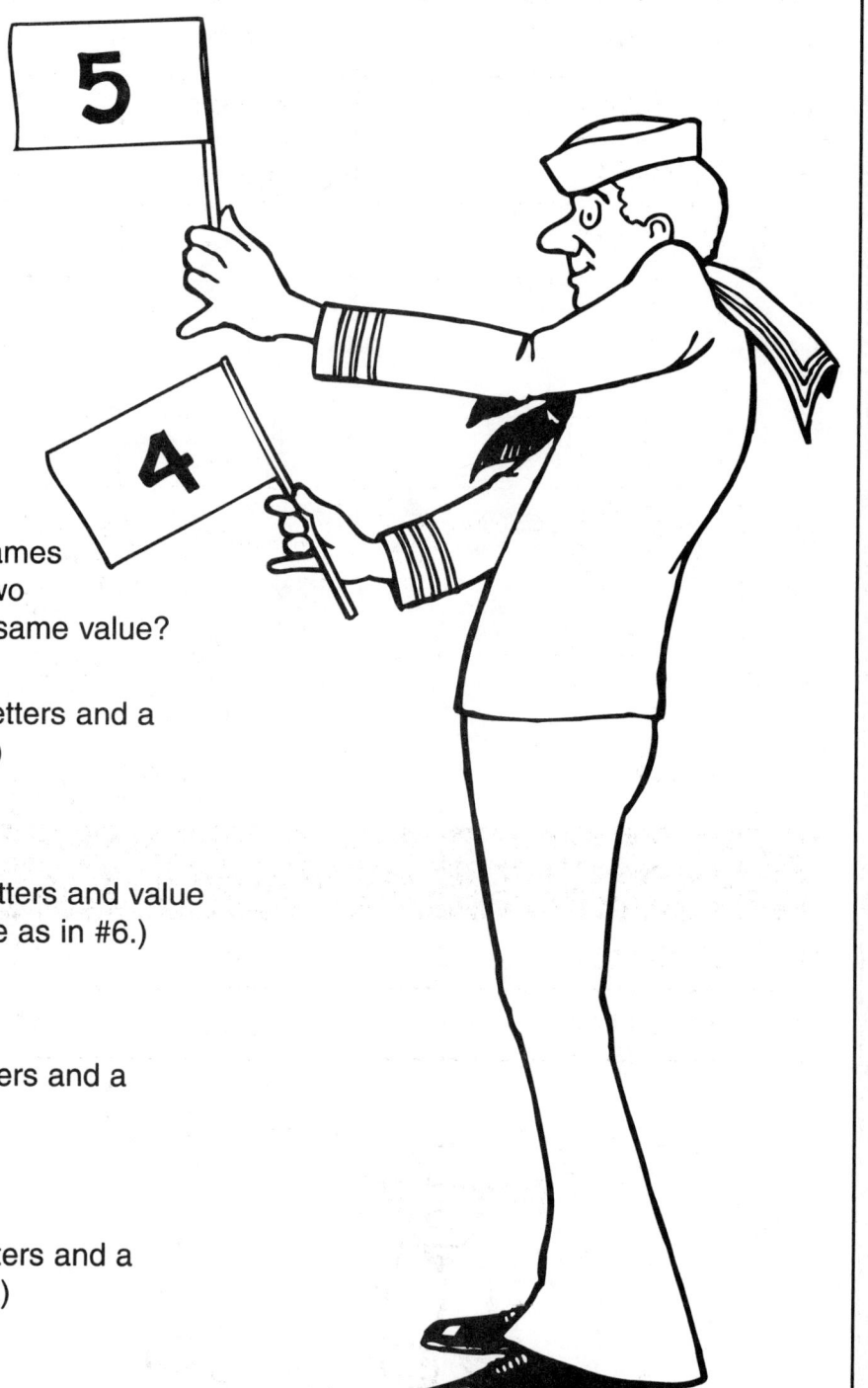

   _____

Name_____

Ordering by Mail

What a great deal! You won a shopping spree in your favorite catalog! Have fun and fill in the missing numbers in the charts below to be sure your order is correct.

	Catalog #	Quantity	Description	Color Number	Size	Price per Item	Total Price
1.	X23A7924	3	jeans	12	10	$16.00	$48.00
2.	Y81F2604	4	sweater	37	M	$23.00	
3.	A12B1805	2	gloves	16	7	$17.00	
4.	W32A1525	6	socks	17	11	$3.00	
5.	X17B7132	4	tie	5	—	$15.00	
6.	B72AG253	3	shirt	23	M	$21.00	
7.	B72B7924	6	hose	1	—	$2.00	
8.						Total for goods	
						Tax	$13.00
						Postage	$7.00
9.						Total cost	

	Catalog #	Quantity	Description	Color Number	Size	Price per Item	Total Price
10.	Z21P2151	2	belt	12	M	$5.00	
11.	A05M1150	5	shirt	15	L	$12.00	
12.						Total for goods	
						Tax	$5.80
						Postage	$4.20
13.						Total cost	

FS122009 Pre-Algebra Made Simple • © Frank Schaffer Publications, Inc.

Multiplication and problem solving **5**

Name _____

Pay Problems

Learn all about gross pay and net pay by completing the problems below. Follow the example to help you.

Example:
A gas station attendant earns $6.50 an hour and works 38 hours per week. Total weekly deductions amount to $53.60. Find the yearly net pay.

Solution:
1. Find the weekly gross pay.
   $6.50 x 38 = $247.00   earnings per hour x # of hours = weekly gross pay
2. Find the weekly net pay.
   $247.00 − $53.60 = $193.40   gross pay − deductions = net pay
3. Find the yearly net pay.
   $193.40 x 52 = $10,056.80   weekly net pay x 52 (weeks in a year) = yearly net pay

	Hours Worked	Rate per Hour	Gross Pay	Deductions per Week	Net Pay	Yearly Net Pay
1.	27	$7.00		$29.00		
2.	38	$12.00		$87.00		
3.	40	$8.00		$56.00		
4.	38	$15.00		$138.00		
5.	38	$6.00		$36.00		
6.	37	$8.00		$41.00		

7. Mary Ryan earns $10 per hour and works 33 hours per week. Her total weekly deductions amount to $81.00. Find her yearly net pay. _____

8. A worker earns $4.00 per hour and works 30 hours per week. The total weekly deductions amount to $17.00. Find the yearly net pay. _____

9. A computer technician earns $9.00 per hour and works 35 hours per week. His weekly deductions total $74.00. What is his yearly net pay? _____

10. A dental assistant has a yearly net pay of $12,480. His weekly deductions total $40.00. What is his weekly gross pay? _____

Gross pay/Net pay

Name_____

Money Problems

Find the total amount owed and the monthly payment for the problems below. Follow the example to help you.

Example:
The owner of a small business borrows $5000 for 24 months. The finance charge is $400, and the loan is to be repaid in equal monthly payments. Find the amount of each payment.

Solution:
1. Find the total amount owed.
   $5000 + $400 = $5400   amount borrowed + finance charge = total amount owed
2. Find the amount of each monthly payment.
   $5400 ÷ 24 = $225   total amount owed ÷ number of months = monthly payment

	Amount Borrowed	Finance Charge	Total Amount Owed	# of Months to Repay	Monthly Payment
1.	$2000	$136		12	
2.	$1500	$132		12	
3.	$3200	$328		24	
4.	$5000	$868		36	
5.	$3500	$256		12	
6.	$1800	$336		24	
7.	$4500	$444		24	

8. Alfred borrows $2500 and agrees to repay the loan in equal monthly installments over 12 months. Finance charges amount to $475.04. Find the amount of each monthly payment. _____

9. Lois borrows $20,000 and agrees to pay back the loan in equal monthly installments over 10 years. Finance charges amount to $7000. Find the amount of each monthly payment. _____

10. Tania borrows $3000 and agrees to pay back the loan in equal monthly installments over 2½ years. Finance charges total $1387.50. Find the amount of each monthly payment. _____

11. Ed borrows $6000. Finance charges total $4389.60. Ed agrees to repay the loan in equal monthly installments over 45 months. Find the amount of each monthly payment. _____

Finance charges

Hit the Road

Find the number of gallons used and the total cost for the problems below. Follow the example to help you.

Example:
The highway m.p.g. (miles per gallon) estimate for a certain car is 25 miles per hour. About how much will it cost to drive the car 1200 miles in highway driving conditions when gasoline costs $1.50 per gallon?

Solution:
1. Find the number of gallons used.
   1200 ÷ 25 = 48 gallons   number of miles ÷ highway m.p.g. = number of gallons
2. Find the cost.
   48 × $1.50 = $72.00   number of gallons × cost per gallon = total cost

1. The city m.p.g. for a certain car is 23 miles. Suppose that gasoline costs $1.60 per gallon. About how much will it cost to drive the car 920 miles in the city?

2. The city m.p.g. for a certain car is 28 miles. Suppose that gasoline costs $1.50 per gallon. About how much will it cost to drive the car 1036 miles in the city?

3. The highway m.p.g. for a certain car is 36 miles. Suppose that gasoline costs $1.40 per gallon. About how much will it cost to drive the car 3312 miles on the highway?

4. The highway m.p.g. for a certain car is 29 miles. Suppose that gasoline costs $1.55 per gallon. About how much will it cost to drive the car 1827 miles on the highway?

5. The city m.p.g. for a certain car is 27 miles. Suppose gasoline costs $1.45 per gallon. If the total cost is $46.40, how many miles did you drive the car on the highway?

6. The highway m.p.g. for a certain car is 32 miles. Suppose that gasoline costs $1.48 per gallon. If the total cost is $39.96, how many miles did you drive the car on the highway?

7. Suppose that gasoline costs $1.65 per gallon. If you drive the car 1404 miles on the highway and spend a total of $89.10, what is the highway m.p.g. for your car?

8. Suppose that gasoline costs $1.60 per gallon. If you drive the car 1326 miles in the city and spend a total of $124.80, what is the city m.p.g. for your car?

Multiplication, division, and problem solving   FS122009 Pre-Algebra Made Simple • © Frank Schaffer Publications, Inc.

Cut and Complete

Cut out the number boxes at the bottom of the page. Use each box (only once) to complete all of the equations below.

1. 3 • (☐ − ☐) = 15

2. (4 • ☐) + ☐ = 24

3. 4 • (☐ − 6) = 12

4. (3 • ☐) + ☐ = 21

5. 4 + (☐ • ☐) = 16

6. 3 • (9 − ☐) = 24

0 1 2 3 4 5 6 7 8 9

Name_____

Fill in the Blanks

Cut out the number boxes at the bottom of the page. Use each box (only once) to complete all of the equations below.

1. $9^2 + 3 \cdot \boxed{} = 96$

2. $\boxed{}^3 - 30 \div 2 = 49$

3. $2^2 + \boxed{}(3 + 10) = 82$

4. $8^2 \div (\boxed{} - 4) = 16$

5. $[(12 - 4) \cdot 2 + 11] \div 3 = \boxed{}$

6. $110 - [\boxed{} + (4^2 - 9)] = 83$

7. $34 + [(16 + \boxed{}) \div 4] - 16 = 25$

8. $[(6 + 8) \cdot \boxed{} - 12] + 6 = 36$

9. $45 + [6^2 - (12 + 6)] = \boxed{}$

10. $48 - [36 \div (4 + 5)] + \boxed{} = 55$

| 12 | 5 | 11 | 3 | 8 | 20 | 4 | 9 | 63 | 6 |

10 Order of operations

FS122009 Pre-Algebra Made Simple ▪ © Frank Schaffer Publications, Inc.

Name_____

Number Fun

Solve each equation using the number cards at the bottom of the page. All three number cards must be used in each equation.

1. ☐ • ☐ − ☐ = 2

2. ☐ • ☐ − ☐ = 8

3. ☐ • ☐ − ☐ = 22

4. (☐ + ☐) ÷ ☐ = 1

5. (☐ + ☐) ÷ ☐ = 5

6. (☐ + ☐) ÷ ☐ = 2

7. (☐ + ☐) • ☐ = 20

8. (☐ + ☐) • ☐ = 32

9. (☐ + ☐) • ☐ = 36

10. ☐ ÷ (☐ + ☐) = 1

| 2 | 4 | 6 |

FS122009 Pre-Algebra Made Simple ▪ © Frank Schaffer Publications, Inc.

Order of operations 11

Name_____

Back Up!

Take the reverse approach to solving math equations. In each problem below, use three of the four given numbers in the box to solve the equation.

| 2 | 3 | 4 | 6 |

Example: 4 x 3 + 6 = 18

1. _____ x _____ ÷ _____ = 1
2. _____ x _____ − _____ = 5
3. _____ ÷ _____ x _____ = 6
4. _____ x _____ ÷ _____ = 8
5. _____ x _____ − _____ = 14
6. _____ x _____ + _____ = 15

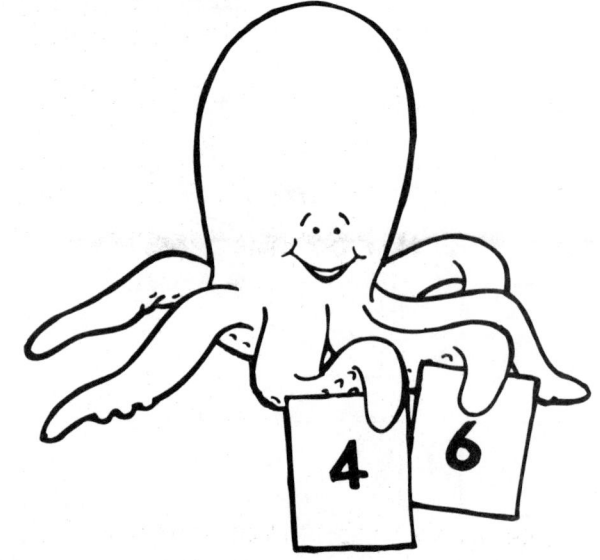

Make up six problems of your own. Trade with a classmate and solve each other's problems. Remember to give the numbers to use in the box. Also provide the operations and the answers.

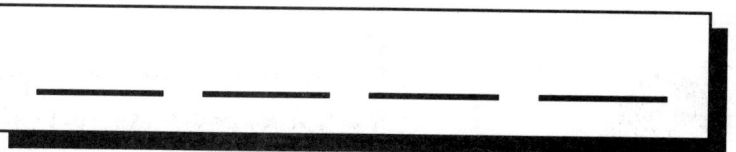

1. _____ _____ _____ _____ = _____
2. _____ _____ _____ _____ = _____
3. _____ _____ _____ _____ = _____
4. _____ _____ _____ _____ = _____
5. _____ _____ _____ _____ = _____
6. _____ _____ _____ _____ = _____

Name_____

Problem solving

Name _____

Three Out of Four

Take the reverse approach to solving math equations. In each problem below, use only three of the four given numbers in the box to solve the equation. You must also insert the operation signs (+, –, ÷, x). Can you come up with more than one equation for a certain answer?

Example: 15 + 3 – 12 = 6

1. _____ _____ _____ = 0
2. _____ _____ _____ = 8
3. _____ _____ _____ = 1
4. _____ _____ _____ = 20
5. _____ _____ _____ = 7
6. _____ _____ _____ = 13

Now make up six problems of your own. Use only addition and subtraction. Trade with a classmate and solve each other's problems. Remember to give the numbers to use in the box and the answers.

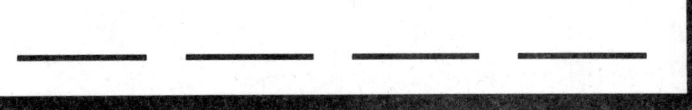

1. _____ _____ _____ _____ = _____
2. _____ _____ _____ _____ = _____
3. _____ _____ _____ _____ = _____
4. _____ _____ _____ _____ = _____
5. _____ _____ _____ _____ = _____
6. _____ _____ _____ _____ = _____

Order of operations

Only Use Half!

Take the reverse approach to solving math equations. Using any two of the four given fractions in the box, write the problems.

Example: $\frac{5}{12} + \frac{11}{12} = \frac{4}{3}$

1. _____ − _____ = $\frac{1}{2}$

2. _____ + _____ = 1

3. _____ − _____ = $\frac{1}{6}$

4. _____ + _____ = $\frac{3}{2}$

5. _____ − _____ = $\frac{1}{3}$

6. _____ + _____ = $\frac{2}{3}$

$\frac{7}{12} \quad \frac{5}{12} \quad \frac{1}{12} \quad \frac{11}{12}$

Challenge: Use all four numbers in the box. _____ + _____ − _____ − _____ = 0

Make up six problems of your own involving fractions. Trade with a classmate and solve each other's problems. Remember to give the numbers to use in the box. Also provide the operations and the answers.

1. _____ _____ = _____

2. _____ _____ = _____

3. _____ _____ = _____

4. _____ _____ = _____

5. _____ _____ = _____

6. _____ _____ = _____

Rows and Columns

Take the reverse approach to solving math equations. Using the numbers 1–9, fill in each puzzle so that each column and each row produces the given answer. (Hint: A number can be used more than once in a puzzle.)

Addition and subtraction

Name_____

More Rows and Columns

Take the reverse approach to solving math equations. Using the numbers 1–9, fill in each puzzle so that each column and each row produces the given answer. (Hint: A number is used more than once in the puzzle at the bottom of the page.)

9	x		+		=	16
x		x		x		
	x		+		=	38
÷		÷		÷		
	x		+		=	10
=		=		=		
15		3		14		

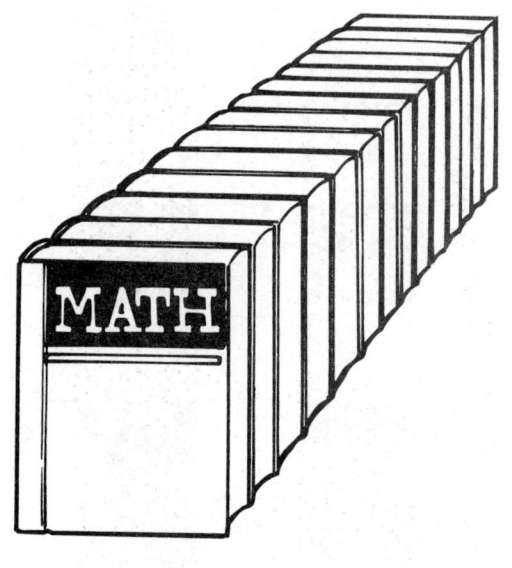

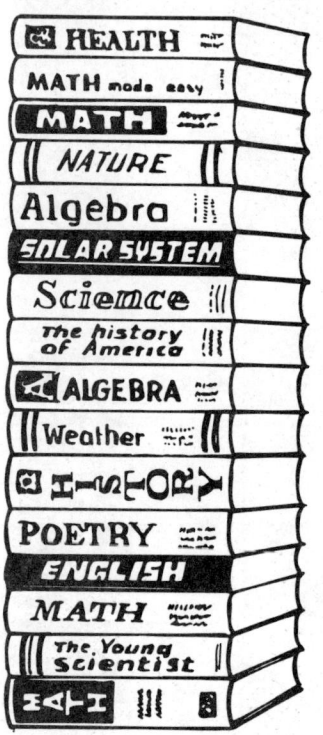

5	x		÷		=	10
x		+		+		
	x		÷		=	18
÷		-		+		
	+		-		=	4
=		=		=		
6		6		13		

16 Addition, subtraction, multiplication, division

Name_____

Adding Every Which Way

Add the numbers below in both directions (rows and columns). Then add the totals. Compare your results in the proof box.

2	4	6	8	3	5	1	7	9	☐
1	3	5	7	9	4	6	2	8	☐
8	9	7	9	7	5	6	8	4	☐
4	6	2	8	3	5	9	5	7	☐
7	5	4	8	3	9	4	7	8	☐
9	7	4	3	5	4	6	7	4	☐
9	7	7	6	4	5	7	9	4	☐
2	5	7	9	6	4	3	7	9	☐
8	6	5	7	4	3	8	6	5	☐
9	8	6	4	5	3	6	8	4	☐
☐	☐	☐	☐	☐	☐	☐	☐	☐	◩ proof box

Just for Fun!
Play the number game by following the directions below.

1. Write your age. _____

2. Multiply by 4. _____

3. Add 10 to your answer. _____

4. Multiply the sum by 25. _____

5. Subtract the number of days in a year. _____

6. Add the change in your pocket. (under $1.00) _____

7. Add 115 to your answer. _____

Addition 17

Let's Compare

A **ratio** is a comparison of two numbers by division. (Like fractions, ratios should be written in simplest form.) For example, if a football team won 6 games and lost 2, the ratio of games won to games lost is $\frac{6}{2} = \frac{3}{1}$.

A **proportion** is a statement that two ratios are equal. For example, $\frac{1}{2} = \frac{2}{4}$, $\frac{2}{4} = \frac{6}{12}$, $\frac{8}{16} = \frac{3}{6}$

Set up a proportion and solve for each problem below. Follow the example to help you.

Example:
A recipe calls for 1 cup of sugar for every 3 cups of flour. If a cook is using 7 cups of flour, how many cups of sugar are needed?

Solution:
1. Write two equal ratios. Let x equal the number of cups of sugar needed.
   $\frac{\text{sugar}}{\text{flour}} = \frac{1}{3} = \frac{x}{7}$
2. Cross multiply and solve.
   $3x = 7 \cdot 1$
   $x = \frac{7}{3}$
   $x = 2\frac{1}{3}$ cups of sugar

1. The scale on a road map is 1 inch = 80 miles. If the distance between two towns measures 1¾ inches, about how many miles apart are the towns?

2. If oranges cost $1.08 a dozen, how much do 8 oranges cost?

3. Concrete is to be mixed in a cement-to-sand ratio of 1 shovel to 4 shovels. How many shovels of cement should be mixed with 60 shovels of sand?

4. A case of soda is $4.99. How much does one can of soda cost?

Ratio and proportion

Problems With Percents

To solve percent problems, the proportion formula below is used.

$$\frac{\%}{100} = \frac{\text{part}}{\text{whole}}$$

Solve the problems by following the example below.

Example:
12 is 75% of what number?
↑   ↑   ↑
part   percent   whole

Solution:
Set up the proportion.

$$\frac{75}{100} = \frac{12}{n}$$

Solve.

$$\frac{75n}{75} = \frac{1200}{75}$$

$$n = 16$$

1. 9 is what percent of 45?

2. 36 is what percent of 40?

3. What is 57% of 900?

4. What is 9% of 98?

5. 80% of what number is 60?

6. 19% of what number is 57?

7. The rent on a community center building increased from $200 a month to $225 a month. What was the percent of increase?

8. Of 200,000 eggs at a fishery, 65% are not expected to hatch. How many eggs are not expected to hatch?

Name_____

Proportional Problems

To solve percent problems, the proportion formula below is used.

$$\frac{\%}{100} = \frac{part}{whole}$$

Set up a proportion and solve for each problem below. Use the example to help you.

Example:
What was the original price?
Save 15% on men's black leather oxfords.
Sale price: $19.95

Solution:
Original price (100%) − savings (15%) = sale price (85%)
$19.95 is 85% of what number?

$$\frac{85}{100} = \frac{19.95}{n}$$

$$85n = 1995$$
$$n = \$23.47$$

How much money do you save if you buy the sale items listed below?

1. Original price: $17.50
   Save 44%

2. Original price: $420
   Save 41%

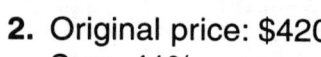

What percent of the original price are you paying for each?

3. Sale price: $1495.00
   Regular: $1649.00

4. Sale price: $57
   Regular: $119

What was the original price of each?

5. Save 19%
   Sale price: $19.96

6. Save 33%
   Sale price: $4.97

Name_____

What Time Is It?

A. Solve the problems below involving time.

1. 2 a.m. + 7 hours and 15 minutes

2. 11:15 p.m. – 5 hours

3. 9:25 a.m. + 3 hours

4. Noon – 2 hours and 15 minutes

5. 3 p.m. – 5 hours and 45 minutes

B. Convert the times in hours:minutes:seconds in the first column in the chart below to minutes. Follow the example to help you.

	Time	Hours	Hours to Minutes	Minutes	Seconds	Seconds to Minutes	Total Time in Minutes
Ex.	2:17:34	2 • 60	120	17	34 ÷ 60	.57	137.57
6.	3:31:22						
7.	4:16:12						
8.	2:26:31						
9.	2:34:20						
10.	2:41:72						
11.	2:16:21						
12.	2:10:43						
13.	2:08:60						
14.	2:04:43						
15.	2:04:08						

Money Matters

Solve the problems below.

A. The balance in Erin's checking account was $295.00. Then she wrote some checks.

1. Find the balance of the checking account after each check was written.

Check	Amount	Balance
225	$8.56	
226	$33.94	
227	$18.39	
228	$29.00	
229	$19.98	
230	$36.29	

2. Make a graph of your choice to display the data.

B. The conversion rate in Mexico during a a recent year in June was 2850 pesos to $1.00. Last June, Wendy and her husband, Bill, went to Cancun, Mexico, to celebrate their third wedding anniversary. Wendy was standing on a beach when a guy walked up selling T-shirts. The price was one T-shirt for 18,525 pesos or three T-shirts for 50,872.5 pesos.

1. Which is the better deal in dollars? Prove your answer.

2. How much will you save if you choose the better deal?

3. Wendy had $40.00. She bought a soda for 5700 pesos and wants to buy 6 T-shirts. What is the cost of 6 T-shirts in pesos and dollars?

4. How much money does Wendy have left in pesos and dollars after buying the soda and 6 T-shirts?

Name_____

More Money Matters

Look up the current rate of U.S. exchange for five countries. (If using the Internet, the site is http://users.hunterlink.net.au/~ddpay/currency/)

Example:

Amount	From	Rate of Exchange	To	Equals
$1	U.S. dollars	1 to 10.09	Mexican pesos	10.09 pesos
$1	U.S. dollars	1 to 134.78	Japanese yen	134.78 yen
$1	U.S. dollars	1 to 1	Bahamian dollars	1 Bahamian dollar
$1	U.S. dollars	1 to 1.727	German marks	1.727 German marks
$1	U.S. dollars	1 to 1.717	Australian dollars	1.717 Australian dollars

Amount	From	Rate of Exchange	To	Equals
$1	U.S. dollars			
$1	U.S. dollars			
$1	U.S. dollars			
$1	U.S. dollars			
$1	U.S. dollars			

1. Calculate the cost of a cheeseburger meal in each country. (U.S. price $3.17)

2. Calculate the cost of a Mark McGwire T-shirt in each country. (U.S. price $15.99)

3. Calculate the cost of a can of soda in each country. (U.S. price $.85)

4. List the pros and cons of making a purchase in a foreign currency on the back of this page.

5. Imagine you have $1500 in U.S. dollars. Make 5 conversions to make the greatest profit through exchange.

Money conversion

Name_____

Making Money

Use your problem-solving skills to answer the problems below.

A. The price of business cards are as follows: Store A—$30.00 for the first 200 cards and $0.05 for each additional card; Store B—$0.10 for each card.

  1. For what number of cards is the cost the same at both stores?

  2. How many cards would you need to order so that store A has the best price?

  3. Is there ever a situation in which ordering less cards than those required for the break-even point might be justified? Write a paragraph supporting your answer.

B. A heavy equipment operator makes an hourly wage of $15.00 per hour, and a laborer makes an hourly wage of $10.00 per hour. The laborer starts preparing for work at 7:00 a.m. Each worker gets a 1-hour (not paid) lunch break. The heavy equipment operator begins at 7:30 a.m.

  1. Explore ways to determine the time of day each worker will have earned the same amount of wages for the day.

  2. How much money will each worker have made at the end of 8 hours?

  3. How much money will each worker have made by 5:00 p.m.? (Workers receive 1½ times their hourly wage for any time worked after 8 hours in a day.)

Problem solving with integers   FS122009 Pre-Algebra Made Simple ▪ © Frank Schaffer Publications, Inc.

Name_____

Ticket Time

Using the chart to the right, answer the questions below.

1. Fill in the table below to show the unit cost per ticket for adults and the unit cost per ticket for youth under 18. (This does not include sales tax.)

# of Tickets	Discount per ticket	Adult unit cost per ticket	Youth unit cost per ticket
10 to 20	$1.50	$6.50	$4.50
21 to 40			
41 to 60			
61 to 80			
81 to 100			
100+			

# of tickets	Discount per ticket
10 to 20	$1.50
21 to 40	$1.75
41 to 60	$2.00
61 to 80	$3.50
81 to 100	$3.75
100+	$4.00

Individual Adult Ticket $8.00

Individual Youth Ticket $6.00

2. Mr. Shelton and 10 teachers attended a Sunday afternoon game. Compute the cost of all tickets, including sales tax. (Use 6.25% for sales tax.) _____

3. The group had so much fun that they decided to each invite 4 more friends to the next game. What was the cost each person paid for a ticket? (Assume all were adults.) _____ What is the cost of all tickets including sales tax? _____

4. What would be the adult cost per ticket if 15 members of the group in problem 3 are under 18? _____ Calculate the cost of the 15 youth tickets (including sales tax). _____ What is the cost of all tickets (adult and youth) including sales tax? _____

5. The Hendersonville Youth Soccer League took more than 100 young people to the last home game. Calculate the cost of each youth ticket, including sales tax. _____

6. Write a proportion to find the percent discount for each of these group tickets based on an adult individual game ticket.

    a. 10–20 tickets _____   b. 21–40 _____   c. 41–60 _____

    d. 61–80 _____   e. 81–100 _____   f. 100+ _____

7. Calculate the percent discount for each of the group tickets using the youth individual ticket price.

    a. 10–20 tickets _____   b. 21–40 _____   c. 41–60 _____

    d. 61–80 _____   e. 81–100 _____   f. 100+ _____

8. What is the range of discounts for adult tickets? _____ for youth tickets? _____

Name _____

What Can We Afford?

Suppose you can spend $3.25 per student in your class and there are 17 students in your class. Let's say the class wants to get T-shirts, posters, pencils, and have a pizza party. Look at the chart below to see what the class can afford. Then answer the questions.

White T-shirts with one design	5 for $12.20	1 for $2.00
Purple T-shirts with design on front and back	10 for $24.65	1 for $2.50
White T-shirts with design on front and back	9 for $21.43	1 for $2.25
posters	3 for $9.99	1 for $4.50
pencils	50 for $3.57	1000 for $9.75
Pizza 1	Each 2-topping pizza is $7.99.	Each one-topping pizza is $6.99.
Pizza 2	First pizza is $14.55.	Each additional pizza is $5.00.

1. Figure out your total budget.

2. Prioritize the items the class wants.

3. Figure out what the class can afford. Explain all the different options.

Calculating cost/Comparisons

FS122009 Pre-Algebra Made Simple • © Frank Schaffer Publications, Inc.

All the Same

For each problem, insert the proper symbols (+, −, x, ÷) in the boxes to make the expression on the left equal to the number on the right.

1. 5 ☐ 3 ☐ 6 = 23

2. 5 ☐ 7 ☐ 2 ☐ 2 = 39

3. 8 ☐ 4 ☐ 1 = 5

4. $\dfrac{24 \;\square\; 16}{15 \;\square\; 10} = 8$

5. 3^2 ☐ [(10 ☐ 5) ☐ 2] = 9

6. (17 ☐ 2) ☐ 10 ☐ 8 = 142

7. (8^2 ☐ 12) ☐ 3 ☐ 3 ☐ 4 = 304

8. [(9 ☐ 5) ☐ 2 ☐ 16] ☐ 5 = 7

9. [75 ☐ 21 ☐ (5 ☐ 2) ☐ 3^2] ☐ 5 = 60

10. 4^2 ☐ 7 ☐ (12 ☐ 3) = 121

Order of operations

Solving Equations

Solving equations and understanding the concept of the "unknown" in an algebraic sentence is a large part of algebra. Students will greatly benefit from the activities involving solving equations in this section. Allow students ample opportunity to work with manipulatives and time to complete several examples with your guidance. Be sure students gain a conceptual understanding of the concepts to the right before proceeding through the student activity pages (pages 30–36).

Present everyday situations to your students in which they may use their new skills. Also relate the learning process to real examples from the students' lives.

CONCEPTS

The ideas and activities presented in this section will help students explore the following concepts:
- Distributive Property
- combining like terms
- solving equations
- exponents
- LCM and GCF

GETTING STARTED

- Have students define the words below.

    distribute   like   terms   expressions   equations   combine   solutions

- To help students visualize combining like terms, put the following manipulatives on a table: pennies, linking cubes, paper clips, pencils (or anything else that is available). Divide the manipulatives into 3–4 groups. Make sure each type of manipulative is repeated in each group. Tell students to combine like manipulatives. Explain how pencils can only be combined with pencils, etc. Expressions can also be written to represent the various groups and the combining of the groups.

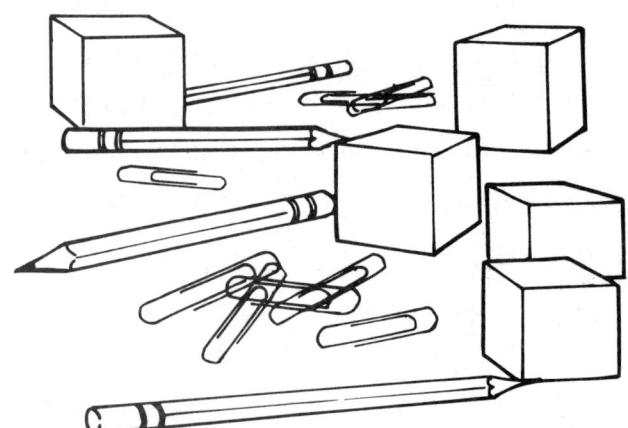

FUN WITH MATH

Introduce students to the concept of distribution by putting the problem below on the board and going over it as a class. To help students get started, let f represent *fries* and s represent *soda*.

You work at McDonalds™ as the manager. You must coordinate the orders below by "distributing" fries and a soda to each order.

Register 1

Register 2

Register 3

Write an algebraic expression to represent the above distribution.

[f = fries, s = soda; $fs(c + m + s) = fsc + fsm + fss$]

FS122009 Pre-Algebra Made Simple ▪ © Frank Schaffer Publications, Inc.

DEMONSTRATING MATH IDEAS

How many different solutions can you come up with by inserting grouping symbols? (Answers will vary.)

1. $30 + 24 \div 6 \cdot 2$
2. $7 \cdot 3^2 - 5 + 6$
3. $11 + 3 \cdot 19 + 2$
4. $14 + 3^3 - 7$

Chapter Group Projects

ON-LINE SCAVENGER HUNT

Divide students into groups of 2–4. Have groups use the Internet to solve the problems below. Groups should have 2–3 weeks with a half-hour of on-line time each day to complete this activity. Groups should keep all of its strategies, notes, etc., in a folder. This activity is evaluated on the basis of student performance. Upon completion of the activities, the groups should prepare oral presentations to share its results with other students.

1. Identify and describe prime numbers and the history associated with them.
2. Interpret data from countries around the world.
3. Describe a female contributor to the field of mathematics.
4. Calculate exchange rates of ten different countries.
5. Calculate and predict trends in the U.S. National Debt.

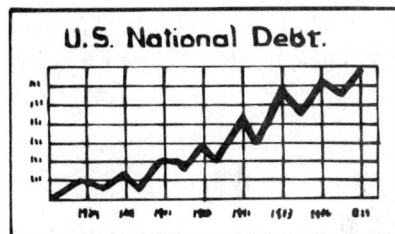

INVESTIGATIVE PROJECT

Recycling one copy of the Sunday newspaper saves +.043 trees. Therefore, recycling the Sunday newspaper for one year saves $52(0.43) = 2.2$ trees.

Have groups of students find five things people can do to protect the environment that can be expressed as positive net daily or annual changes. Have students calculate the expected net changes by 2020. Groups can put their information in a chart like the one below.

Description of gain	Net daily or annual change	Net change by 2020
1		
2		
3		
4		
5		

Follow-up activity:
Have students investigate three items the school recycles and estimate how much money the school's recycling saves.

Sweet Expressions

M&M™'s can be used to represent mathematical expressions. Let the green M&M™'s represent the variable *n*, and the red M&M™'s represent the number 1. Using this system, the expression 3n + 4 would be represented as shown below.

Use this same format to complete Activities I and II below.

Activity I

1. Get a bag of green and red M&M™'s from your teacher.

2. How many green M&M™'s are there in all? _____ How many red M&M™'s are there in all? _____

3. Use your answer to Step 2 to complete the expression below so that it represents the total value of your red and green M&M™'s.

    ___n + ___

4. Can you create subgroups of green and red M&M™'s that are equivalent?

Activity II

1. Place green and red M&M™'s on your desk to represent 6n + 12.

2. Rearrange the M&M™'s in Step 1 into the number of groups indicated below. Then complete each expression so that it represents the total value of M&M™'s in Step 1.

    a. two groups 2(___n + ___)   **b.** three groups 3(___n + ___)   **c.** six groups 6(___n + ___)

3. What can you conclude about the relationship between each expression in Step 2 and the expression in Step 1?

4. Write three expressions to represent the total value of M&M™'s shown in the expression below.

    18n + 24

Distributive Property

Figure "Eat" Out

Complete the activities using M&M™'s. Use the variables below in your expressions.

y = yellow r = red b = brown g = green o = orange

1. Place 3 green M&M™'s in a pile. Make another pile of 5 orange M&M™'s and a third pile of 6 brown M&M™'s.

    a. Write an expression to represent the three piles.

    b. If you eat 2 green and 3 brown candies, what is your new expression?

2. Place 5 brown M&M™'s in a pile, 7 red in another pile, 3 brown in another pile, and 8 green in another pile.

    a. Write an expression to represent the four piles.

    b. Can you combine any of these piles? What would your new expression be?

    c. If you eat 6 brown, 3 red, and 8 green candies, what is your new expression?

3. How would you represent a pile of 5 green and 5 red M&M™'s?

4. Make 5 piles of M&M™'s, any combination you want.

    a. Write an expression to represent your 5 piles.

    b. Eat 6 M&M™'s and write a new expression.

    c. Eat 4 more M&M™'s, all different colors (if possible) and write a new expression.

5. Using all the different colors of M&M™'s, make up an expression.

    a. Eat 4 M&M™'s. What is your new expression?

6. How can you represent peanut or almond M&M™'s? Are they the same as plain M&M™'s?

7. Place 7 yellow M&M™'s in a pile. Make another pile of 5 orange and a third pile of 6 brown M&M™'s.

    a. Write an expression to represent the three piles.

    b. If you eat 2 yellow and 12 brown, what is your new expression?

Combining like terms

Make It Simple!

Simplify each expression.

1. $a^5 \cdot a^4$
2. $m^7 \cdot m^4$
3. $x^8 \cdot x^6$

4. $b \cdot b^7 \cdot b^6$
5. $y^3 \cdot y^9 \cdot y^5$
6. $(10n^2)(3n^3)$

7. $7x^3 \cdot 3y^2$
8. $7d^8 \cdot 4d^7$
9. $(5x^3)(6x^5)$

10. $(9m^3)(4m^5)$
11. $(3a^4)(9x^3)$
12. $a^{10} \cdot a^3 \cdot a^4$

13. $8a^9 \cdot 7a^3$
14. $(6b^3)(7b^6)$
15. $5m^{11} \cdot 9m^4$

16. $(3c)(9c^8)$
17. $10a \cdot 3a^{14}$
18. $(4a)(3a^6)(7a^4)$

19. $(6c)(7d)(3c^2)$
20. $(8x)(5y)(7x)$
21. $(5c^2)(4d)$

22. $(6x^7)(5x^3)(9x^8)$
23. $(7c^3)(3d^2)(c^5)$
24. $(5m^5)(4x^3)(8m^4)$

25. $4a^5 \cdot 10a \cdot 6a^3$
26. $6d^3 \cdot 8c^4 \cdot 10d^7$
27. $(9w^2)(2w^3)(5y^3)$

28. $a^7 \cdot b^3 \cdot b^5 \cdot a$
29. $m^6 \cdot m^7 \cdot n^3 \cdot n^{12}$
30. $x^7 \cdot x^4 \cdot x \cdot x^{10}$

31. $(4a^3)(3x^2)(10x^4)$
32. $2d^3 \cdot 3a^2 \cdot 5d^7 \cdot 8a^4$
33. $6x^5 \cdot 10y \cdot x^7 \cdot 3y^5$

Combining like terms

Super Scales

Balance the scales by putting numbers in the polygons. In each problem, polygons that are the same must have the same numbers. There are many possible answers for each scale problem. Find three possible solutions for each problem.

   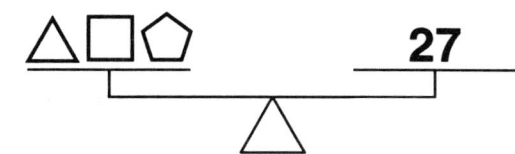

1. ____ + ____ + ____ = 24   2. ____ + ____ + ____ = 27
   ____ + ____ + ____ = 24   ____ + ____ + ____ = 27
   ____ + ____ + ____ = 24   ____ + ____ + ____ = 27

3. ____ + ____ = 16   4. ____ + ____ + ____ + ____ = 32
   ____ + ____ = 16   ____ + ____ + ____ + ____ = 32
   ____ + ____ = 16   ____ + ____ + ____ + ____ = 32

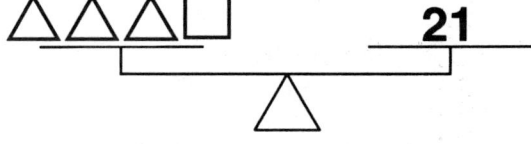   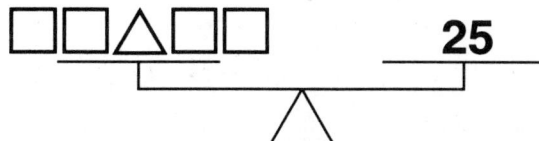

5. ____ + ____ + ____ + ____ = 21   6. ____ + ____ + ____ + ____ + ____ = 25
   ____ + ____ + ____ + ____ = 21   ____ + ____ + ____ + ____ + ____ = 25
   ____ + ____ + ____ + ____ = 21   ____ + ____ + ____ + ____ + ____ = 25

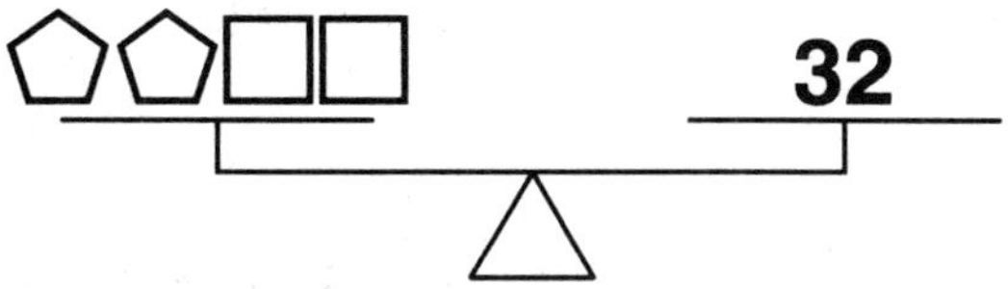

7. ____ + ____ + ____ = 12   8. ____ + ____ + ____ + ____ + ____ = 30
   ____ + ____ + ____ = 12   ____ + ____ + ____ + ____ + ____ = 30
   ____ + ____ + ____ = 12   ____ + ____ + ____ + ____ + ____ = 30

Solving equations

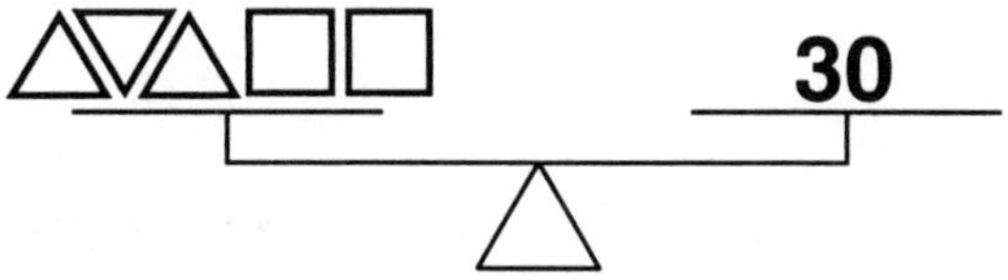

A Sinking Ship

In 1985, 73 years after its maiden voyage, the wreckage of a luxury liner was found in the Atlantic Ocean southeast of Newfoundland. What is the name of this famous ship?

To find out, solve the equations below. Write the letter of the problem above its solution at the bottom of the page. (Hint: Some letters will not be used.)

A. $87 + x = 96$

E. $x/6 = 90$

T. $x/8900 = 0$

R. $2x = {-}90$

I. ${-}11 = x - 8$

G. $x + 56 = 458$

C. ${-}15x = 15{,}000$

H. ${-}54 = x - 23$

N. $x + 5 = 19$

I. $x/5 = 12$

T. $x - 42 = 0$

T. $x + 80 = {-}130$

S. $x + 18 = 22$

D. $29 + x = {-}36$

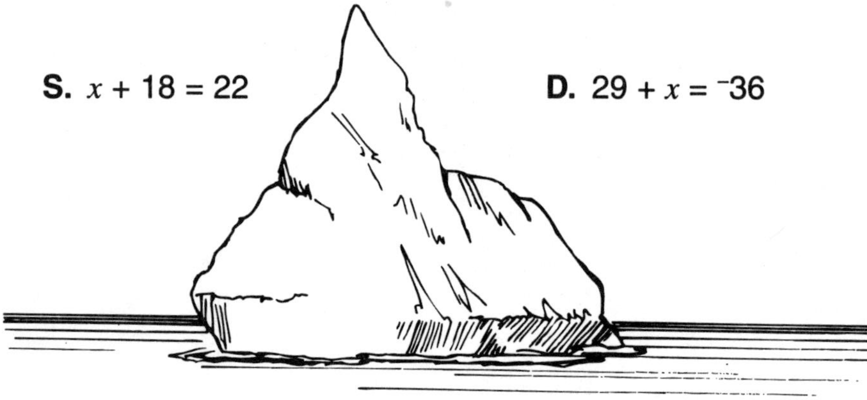

___ ___ ___   ___ ___ ___ ___ ___ ___ ___
 0   -31 540   -210 -3   42   9   14   60 -1000

Solving equations

Stepping Out

With a partner, solve the equations below. Remember to show each step.

1. $2x + 3x = 25$

2. $5m - 8m = 36$

3. $4(7x - 3) = 16$

4. $25 = 5(2t - 7)$

5. $10n - 8 + 4n = 20$

6. $4a + 8 - a = 80$

7. $15r - 18r = 687$

8. $3(m + 1) + 2m = 88$

9. $7(3k + 4) - 18k = 64$

10. $10 = 10(y - 5) + 5y$

Solving equations

Name_____

Estimate and Calculate

For each problem below, first estimate the least common multiple (LCM) and the greatest common factor (GCF). Then calculate the LCM and GCF. Were your estimates close?

1. 4 and 20
   Estimates: LCM _____, GCF _____
   Calculated: LCM _____, GCF _____

2. 60 and 40
   Estimates: LCM _____, GCF _____
   Calculated: LCM _____, GCF _____

3. 80 and 200
   Estimates: LCM _____, GCF _____
   Calculated: LCM _____, GCF _____

4. $18a$ and $36a$
   Estimates: LCM _____, GCF _____
   Calculated: LCM _____, GCF _____

5. $4x^2$ and $20x$
   Estimates: LCM _____, GCF _____
   Calculated: LCM _____, GCF _____

6. $15m^3$ and $25m^2$
   Estimates: LCM _____, GCF _____
   Calculated: LCM _____, GCF _____

7. 18 and 27
   Estimates: LCM _____, GCF _____
   Calculated: LCM _____, GCF _____

8. 63 and 6
   Estimates: LCM _____, GCF _____
   Calculated: LCM _____, GCF _____

9. 8, 12, and 30
   Estimates: LCM _____, GCF _____
   Calculated: LCM _____, GCF _____

10. $7m$, $14m^2$, and $19m$
    Estimates: LCM _____, GCF _____
    Calculated: LCM _____, GCF _____

LCM and GCF

Geometry

All students will greatly benefit from the activities involving geometry in this section. It is important for students to make connections between algebra and geometry. Be sure students gain a conceptual understanding of the concepts to the right before proceeding through the student pages (pages 39–53).

Geometric and spatial relationships involve measurement (length, area, volume), similarity, and transformations of shapes. Geometry involves the study of visual patterns used in representing and describing the world in which we live. Spatial sense is needed to interpret, understand, and appreciate our geometric world.

CONCEPTS

The ideas and activities presented in this section will help students explore the following concepts:

- perimeter
- area
- surface area
- transformations
- rotations
- lines of symmetry
- diameter
- length
- volume
- scale factor
- reflections
- translations
- circumference

GETTING STARTED

Put the problems below on the board to help students get acquainted with geometry.

Given: The vertices of a rectangle are A (4, -5), B (4, 1), C (1, 1), and D (1, -5).

1. Find the midpoint of each side of the rectangle. [(4, -2), (2.5, 1), (1, -2), (2.5, -5)]
2. Find the length of each side of the rectangle. (6, 3, 6, 3)
3. Find the perimeter of the rectangle. (18 units)
4. Find the area of the rectangle. (18 units)
5. If the perimeter of the rectangle is 530, solve for x and state the area. (length = $5x$, $w = 60$; $x = 41$; area = 12,300)

EXPLORING EXTENSIONS

- Have students use land surveys to calculate area using two different methods. Students should explain, contrast, and compare the two methods.

- Have students investigate the volume of an open-ended box that may be constructed from posterboard, measuring 20 cm by 30 cm. As a class, discuss the method(s) used to find the dimensions, allowing for the maximum volume. Students should justify their solutions.

- Provide students with local phone books. Have groups of students estimate the volume of the phone book. Groups should estimate the total volume of all phone books to be recycled in your school and community. Then students can research the amount of landfill rescued as a result of this recycling project. Comparing and contrasting disposal methods to project the cost benefits to the community and the environment is a great follow-up for this activity.

FUN WITH MATH

- Have students measure the perimeter of the room using only their bodies.
- Improve students' lateral thinking—look at volume another way! While looking at volume, try having students find the volume of different sizes of boxes by filling them with Ping Pong™ balls!

BLOCK LETTERING IN PERSPECTIVE

Perspective drawing techniques can be used to create letters or words that appear to be solid—useful for giving emphasis to an element of design. Have students draw a word with one-point perspective by following the directions below.

1. Write a word in block letters. Draw a horizontal line parallel to the bottom edge of the word. Select a vanishing point on the horizontal line.
2. Draw vanishing lines from each corner point of the block letters back to the vanishing point. Select a thickness for your block letters and draw a line (line l) parallel to the horizontal line.
3. To create the back edges of your letters, draw lines parallel to front edges. Erase all the vanishing lines and shade in all the sides and tops of the letters.

Example:

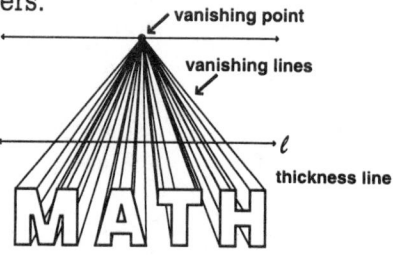

TESSELLATIONS

Use transformations to begin to train the students' eyes to see shapes, to notice the arrangement of those shapes, and to identify patterns. Begin the lesson on tessellations by viewing 12 pictures of transformations. Have students make notes of observations about each picture by comparing and contrasting them. Upon completion, divide students into pairs to compare their observations. Suggest to students that they try grouping the pictures that share similar characteristics. Then match pairs with another pair to group all 12 pictures and list the properties they share. Follow up with a whole group discussion and conclude with all pictures placed in a group with properties listed. Give each group a name (flips, slides, and turns). Then give each student a copy of pages 42–45 and Miras. The pages include introductory and investigative activities to teach students how to do reflections, translations, and rotations using Miras. The transformations they can use to tessellate their figures are reflections (flips), translations (slides), and rotations (turns).

As a final project on tessellations, have each student create a figure (using a 3" x 5" card) and tessellate it on a piece of posterboard. Then students can color their tessellations. Display the completed projects around the room.

Chapter Group Project

Divide students into groups of four. Have each group design a diagram for a candy box that contains five different kinds of candy. Tell students that no two of the same kind of candy can be next to each other on the same row, in the same column, or on the diagonal. Ask students if the candy boxes could be designed in more than one way. Have the groups justify their answers.

Trip Planning

1. Draw a map of your community to scale. Include your house, the school, a zoo, a mall, your favorite restaurant, and one landmark. Don't forget to write the scale for your map.

2. Calculate the distance from your house to the school, to the mall, and then back home.

3. If you average 35 m.p.h. on your trip in #2, how long will your trip take?

Name_____

Soccer Geometry

Learn more about the relationship of soccer and geometry by answering the questions below and on page 41. You may want to use a soccer ball for this activity.

1. What shape is a soccer ball?

2. What polygons are on a soccer ball?

3. Are they regular polygons?

4. How many diagonals can be drawn in each polygon?

5. What is the sum of the angle measures of each polygon?

6. Suppose a soccer ball measures 28 inches in circumference. Calculate the circumference in centimeters.

7. Determine the length of the diameter in inches and centimeters using the formula, $C = \pi d$.

8. What is the length of the radius in inches and centimeters?

9. Determine the surface area of the soccer ball in inches and centimeters using the formula for surface area of a sphere. $A = 4\pi r$

10. Calculate the volume of the soccer ball in inches and centimeters. $V = \frac{4}{3}\pi r$

Geometry

Soccer Geometry continued

11. Soccer field sizes vary from 100 to 130 yards long (touch lines) and 50 to 100 yards wide (goal lines). Convert these measurements to meters.

12. Find the minimum and maximum perimeters of the field in yards. Express the results as an inequality. (minimum < perimeter < maximum)

13. Calculate the minimum and maximum area of the field in yards. Express the results as an inequality. (minimum < area < maximum)

14. Goals are in the center of the goal lines. The distance between the goal posts is 24 feet; these are joined at the top by a crossbar that is 8 feet above ground. Express the distance between the goal posts in yards and meters. Express the height above the ground in meters.

15. The goalie cannot be charged in the goal area, which is a rectangular area 20 yards wide and extending 6 yards in front of the goal. Calculate this rectangular area.

16. The penalty area, in front of the goal, measures 44 yards wide and extends 18 yards in front of the goal. Calculate this rectangular area.

17. The kickoff is from the center spot which is a point in the middle of the halfway line. Opponents must be at least 10 yards away from the ball. What is the circumference of this circle? What is the area of this circle?

18. On a sheet of unlined paper, make a scale drawing of a soccer field. Include goals, goal area, penalty area, halfway line, and kickoff circle.

Mira Mira on the Wall

Use the Miras to reflect the point, line segment, and triangle.

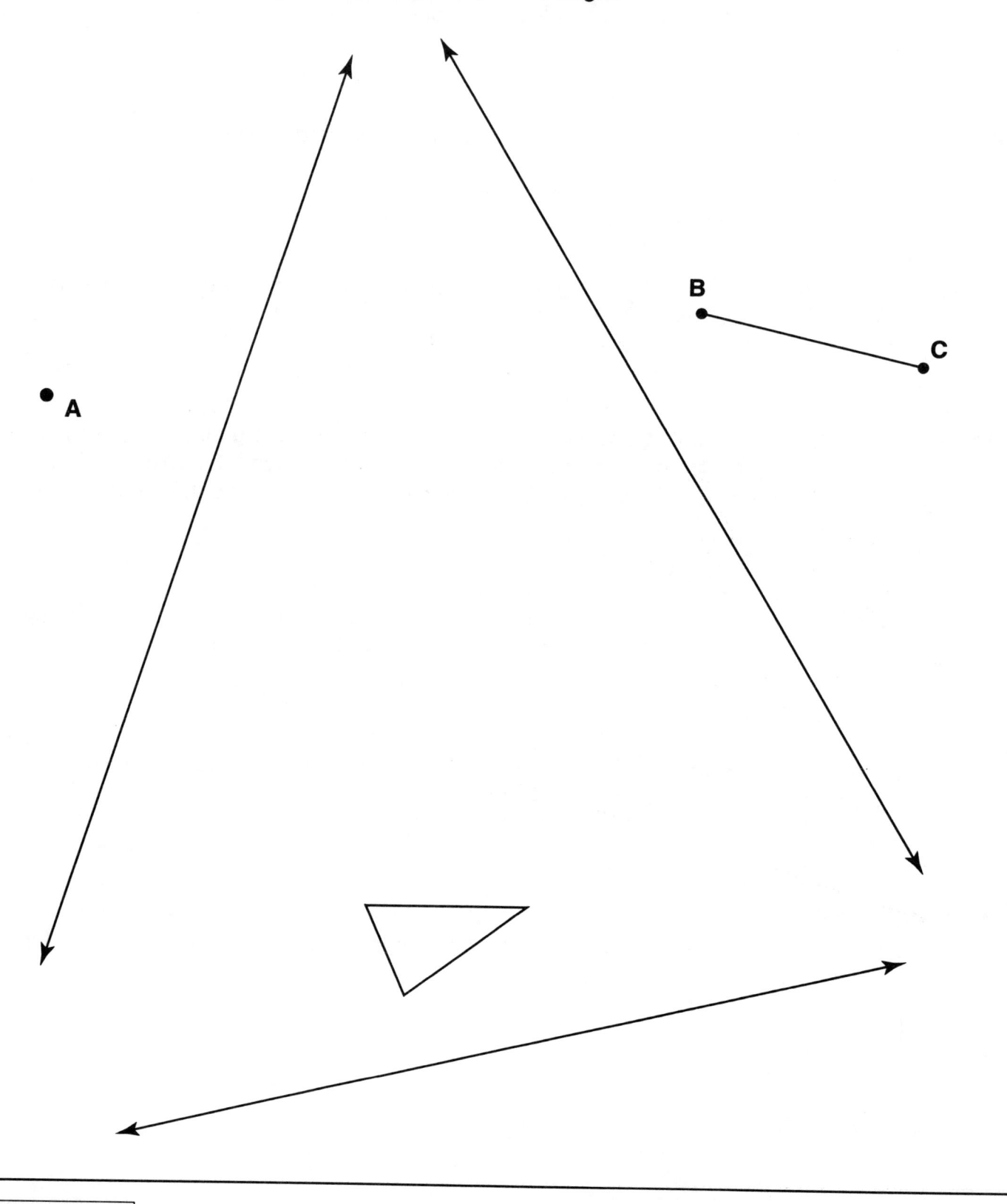

• A

B
•———•C

42 Tessellation

FS122009 Pre-Algebra Made Simple • © Frank Schaffer Publications, Inc.

Reflection Reflection

Use the Miras to translate the point, line segment, and triangle.

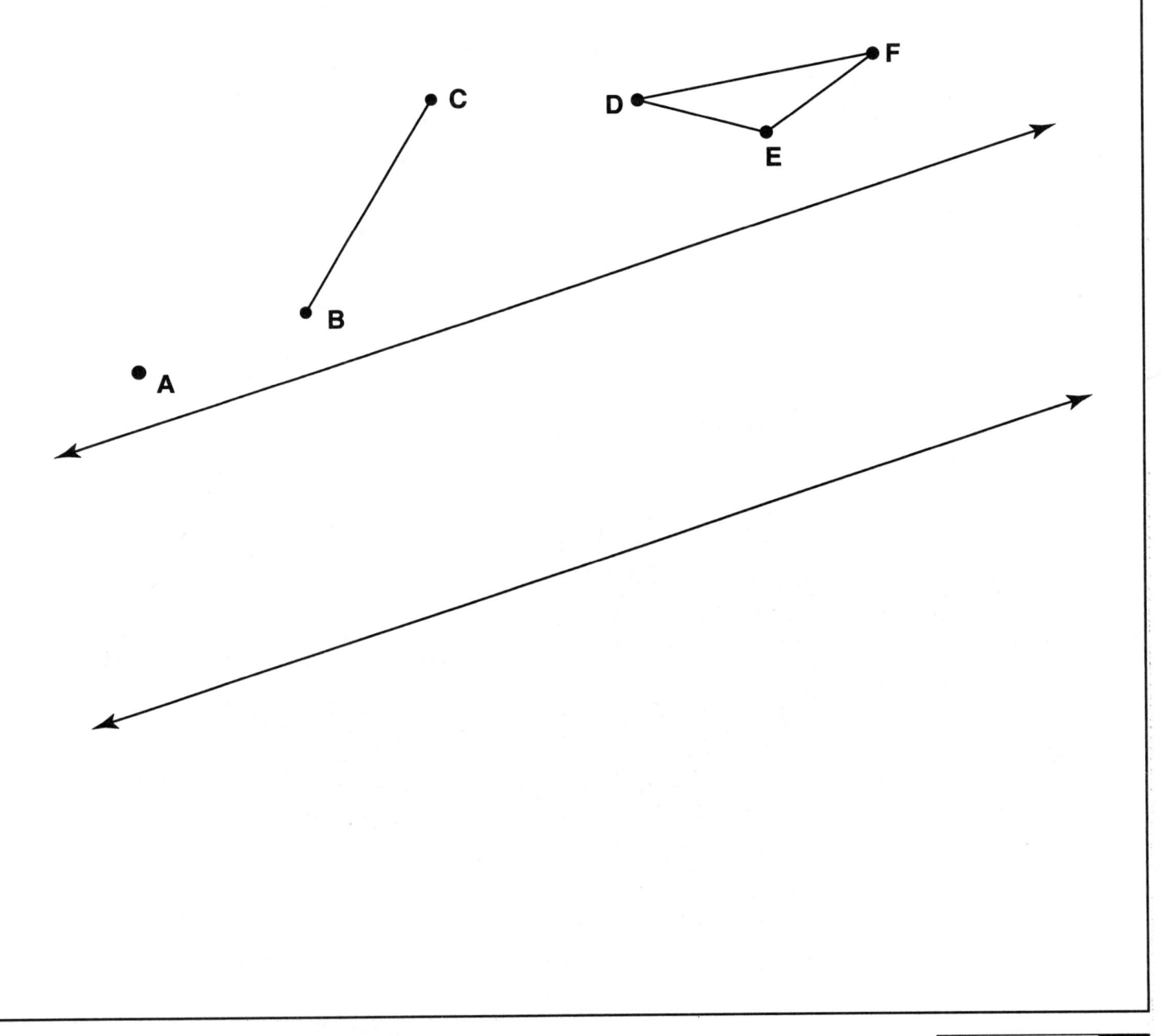

Tessellation 43

Name_____

Rotation Station

Use the Miras to perform the following rotations.

1.

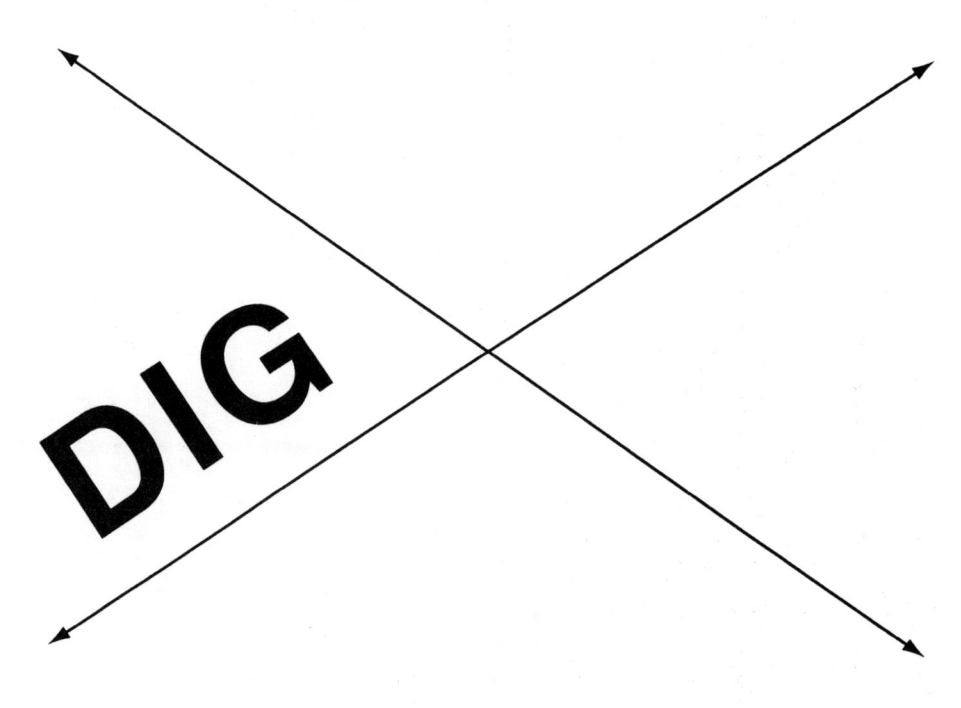

2.

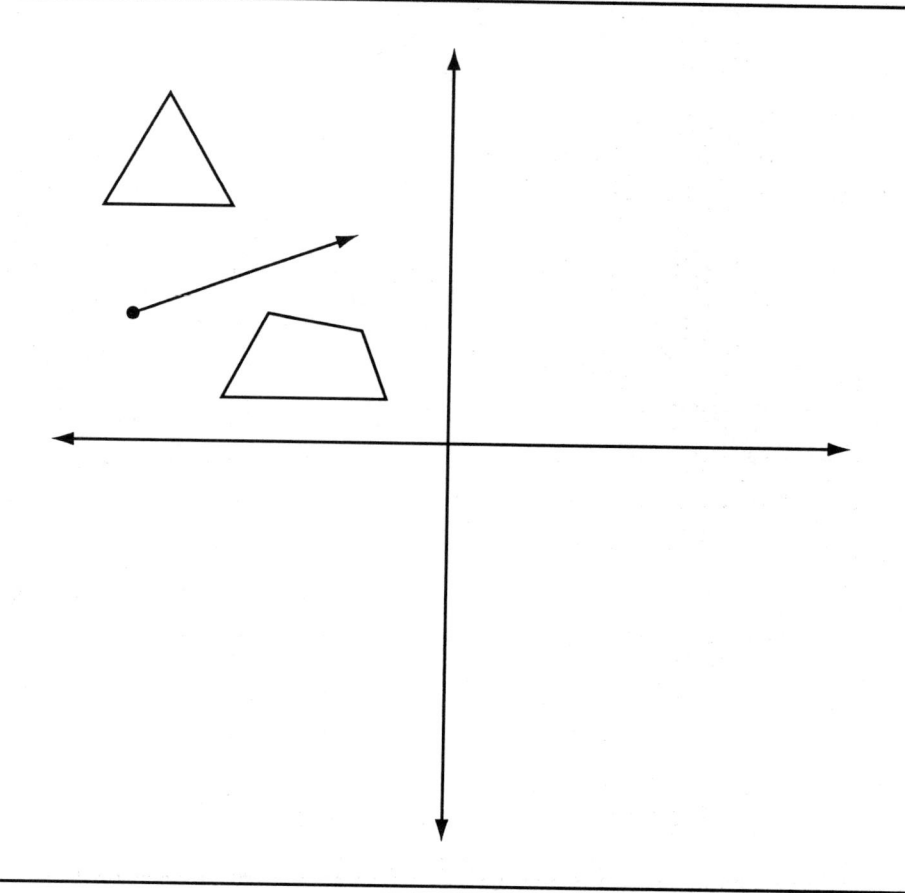

Tessellation

Turn Turn Turn!

Use the Miras to rotate the point, line segment, and triangle.

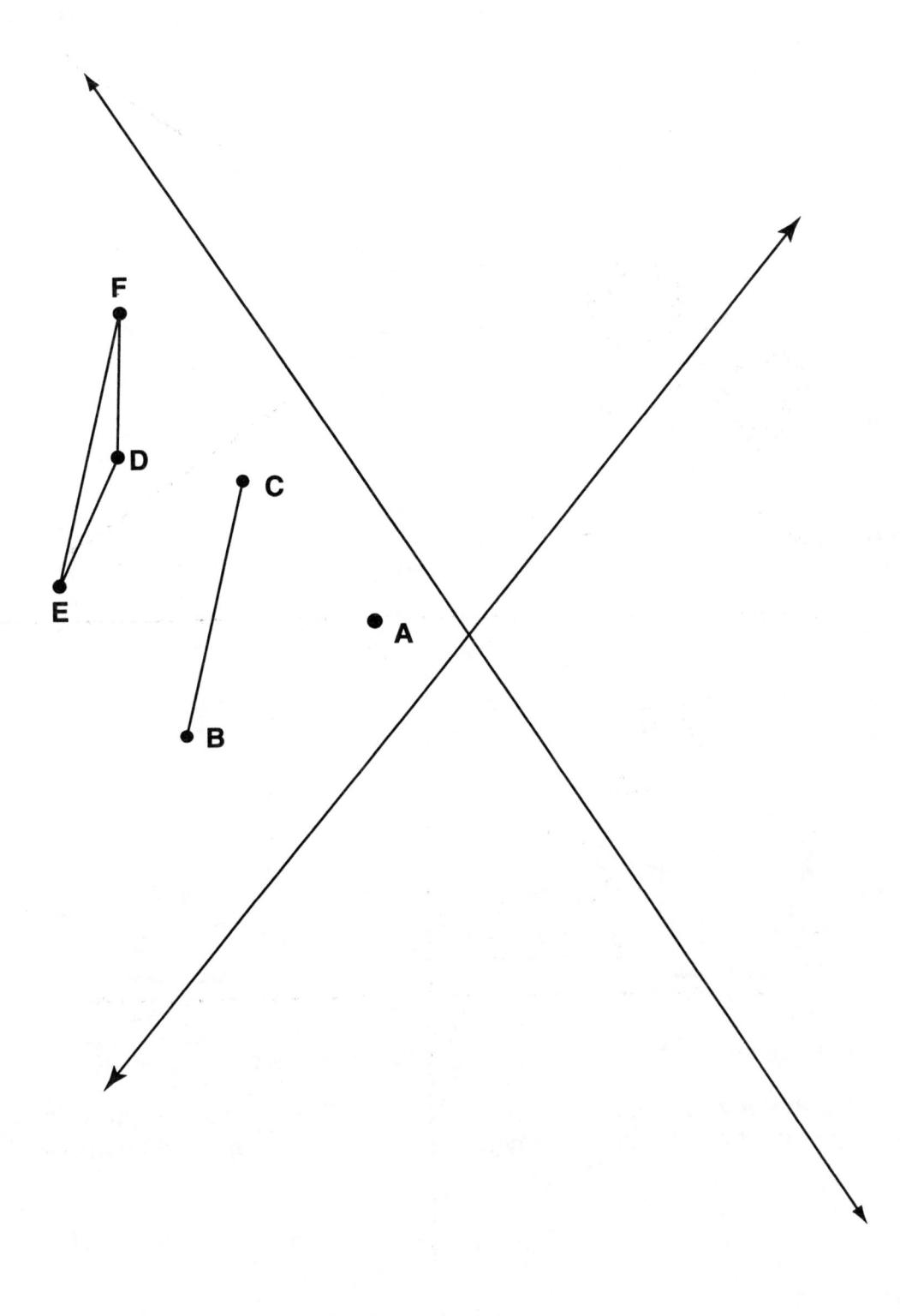

Tessellation 45

Super Simple and Symmetrical

It is possible to fold some figures along a certain line so that one-half of the figure fits exactly over the other half of the figure. Such a line is called a **line of symmetry**. Follow the directions below to learn more about lines of symmetry.

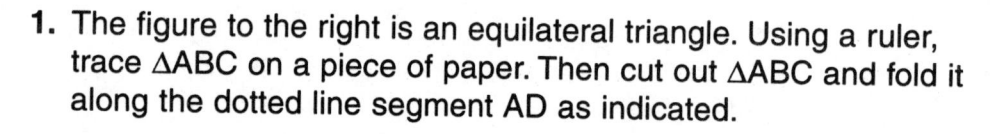

1. The figure to the right is an equilateral triangle. Using a ruler, trace △ABC on a piece of paper. Then cut out △ABC and fold it along the dotted line segment AD as indicated.

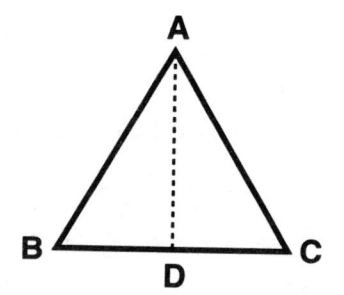

2. Does one-half of the folded triangle in Step 1 fit exactly over the other? _____ Is $\overline{AD}$ a line of symmetry for △ABC? _____

3. Is there another way to fold the triangle in Step 1 so that one-half will fit exactly over the other? _____ Continue folding △ABC to determine how many lines of symmetry it has. _____

4. The figure to the right is a square. Using a ruler, trace the square onto a piece of paper and then cut it out.

5. How many lines of symmetry does the square in Step 4 have?

6. The figures below are regular. Using a ruler, trace and cut out each figure. Then fold each figure to find the number of lines of symmetry it has.

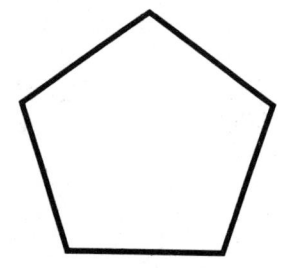

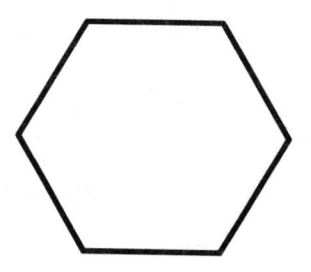

   lines of symmetry _____   lines of symmetry _____

7. Refer to your answers to Steps 3, 5, and 6. Develop a rule for determining the number of lines of symmetry that any regular polygon will have based upon the number of sides it has.

Lines of symmetry

Name_____

Circle Work

The distance around a circle is called its **circumference**. The **diameter** of a circle is the length of a segment passing through the center of the circle with both endpoints on the circle. There is a special relationship between the circumference and the diameter of a circle. To determine what it is, follow the directions below.

1. On another piece of paper, draw a circle with a diameter of 8 centimeters using a compass.

2. Cut out the circle and mark a point on its edge.

3. Roll the circle along a metric ruler for one complete turn, starting and stopping at the mark on the edge. Record the circumference in the table below.

4. Use a calculator to divide the circumference of your circle by its diameter (8 cm). Then complete the table below by repeating Steps 1–4 with circles having the diameters indicated. Round answers in the third column to the nearest hundredth.

diameter (d)	circumference (C)	circumference ÷ diameter
8 cm		
5 cm		
10 cm		
13 cm		

5. Compare your answers in the third column on the chart. What do you notice?

6. Find the mean of the values in the last column of the table in Step 5.

7. Using your answer to Step 6, replace each blank with the correct expression:

$$\frac{C}{d} \approx \underline{\hspace{1cm}} \text{ so } C \approx \underline{\hspace{1cm}}$$

Circumference and diameter

Name _____

Measuring Mania

1. Measure the length and width of each object in the chart to the nearest tenth of a centimeter.
2. Compute the areas and the perimeters. Round each answer to the nearest centimeter.
3. Arrange the areas in order from least to greatest.
4. Arrange the perimeters in order from least to greatest.

Object	Length	Width	Area	Perimeter
door				
chalkboard				
window				
piece of notebook paper				
top of your math book				
side of your math book				
front of teacher's desk				
poster				
floor tile				
side of cabinet				

48 Perimeter and area

FS122009 Pre-Algebra Made Simple ■ © Frank Schaffer Publications, Inc.

Turn Up the Volume!

Learn about volume for cylinders and cones by completing the activity below.

Materials Needed:
a cone-shaped drinking cup
a cylinder with the same radius and height as the cup
metric graduated cylinder
rice
metric ruler
paper and pencil

(The cone and cylinder must be hollow. There must be no base on the cone, and only one base on the cylinder. If a cylinder cannot be found to match the cone, one can be made with cardboard, paper, and a rubber band.)

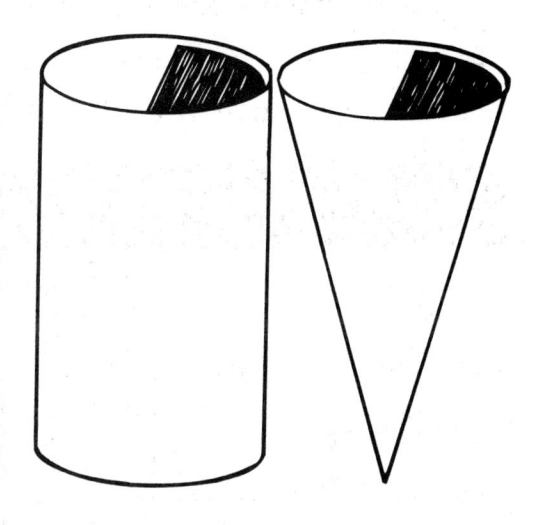

1. Fill the cone to the rim with rice. Then pour the rice into the cylinder. Repeat this process until the cylinder is full. How many full cones of rice were you able to pour into the cylinder?

2. Pour the rice from the full cylinder into the graduated cylinder to find the volume of the cylinder.

3. Predict the volume of the cone based upon your answers to Steps 1 and 2. Fill the cone with rice and pour it into an empty graduated cylinder to check your prediction.

4. Measure the radius of the cylinder and use it to find the area of the base of the cylinder. Then measure the height of the cylinder.

5. Describe the relationship between the area and height found in Step 4 and the volume found in Step 2.

6. Write a formula for the volume of a cylinder based upon the relationship you described in Step 5.

7. Use your answers to Steps 1 and 6 to write a formula for the volume of a cone.

Volumes of cylinders and cones

Awesome Areas

Surface area refers to the area of the entire surface of a space figure. Follow the directions below to learn about surface areas of prisms.

Materials Needed:
scissors   paper and pencil
tape   pattern on page 51
metric ruler

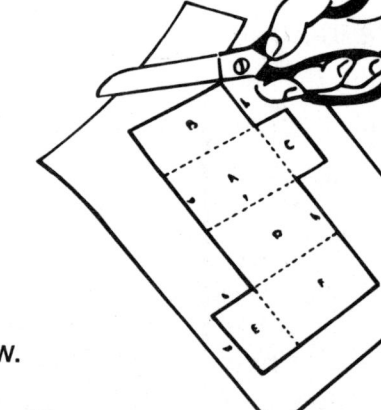

1. Use the figure on page 51 to complete the table below.

Face	Area
A	lw
B	
C	
D	
E	
F	

2. Cut out the figure on page 51. Fold the figure on the dotted lines and tape the edges that meet together. What is the space figure that you have constructed?

3. Which faces of your space figure have the same area?

4. Use your answer to Step 3 to replace each blank with the correct face.

    Surface Area = 2(Area of Face ____) + 2(Area of Face ____) + 2(Area of Face ____)

5. Use your answers to Steps 1 and 4 to write a formula for the surface area of your space figure.

6. Measure the dimensions of your space figure in centimeters. Then replace each blank with the correct measure.

    $l =$ _____ $w =$ _____ $h =$ _____

7. Use your answers to Steps 5 and 6 to find the surface area of your space figure.

Surface area of a prism

Awesome Areas continued

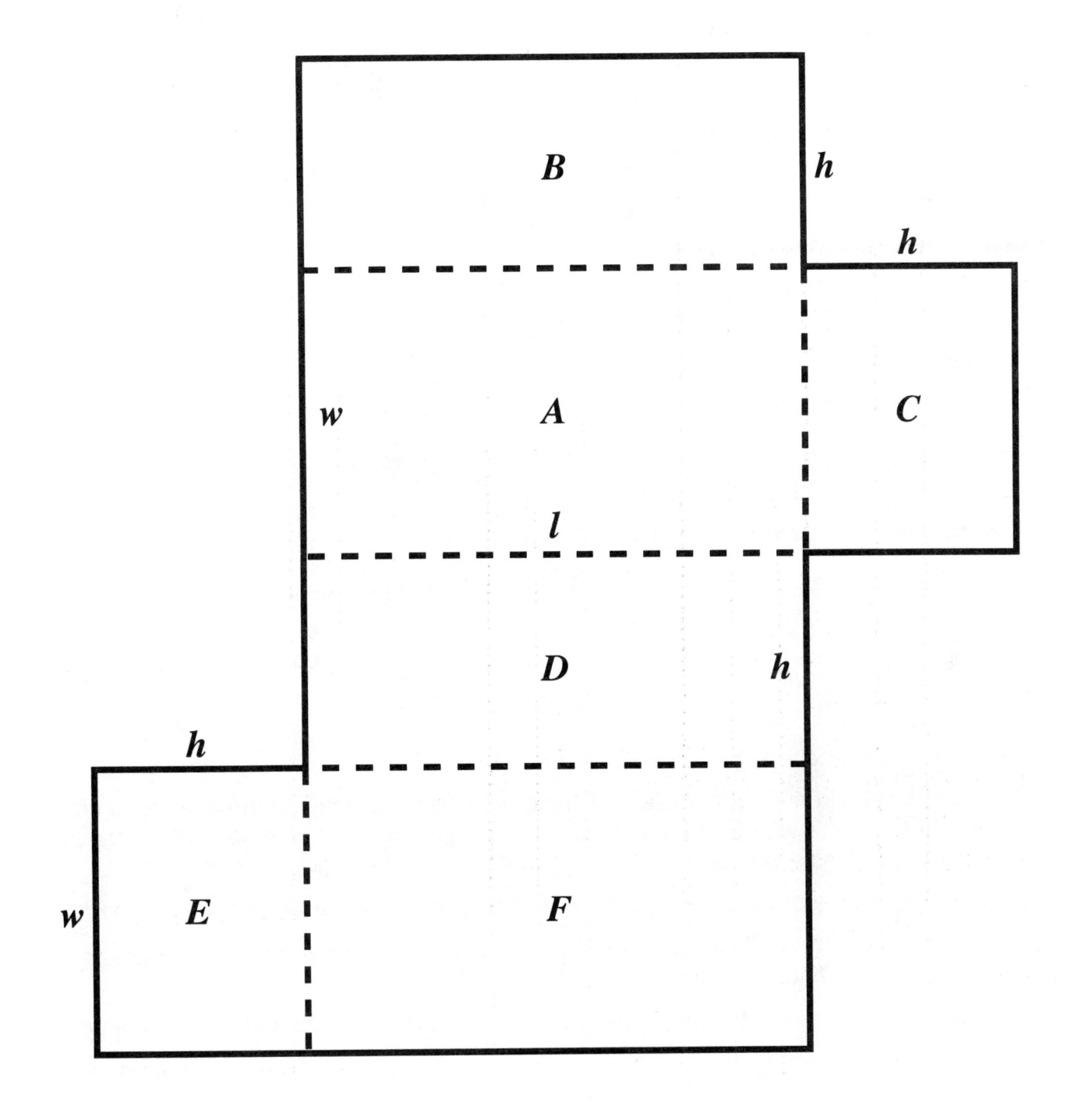

Surface area of a prism

Name_____

Triangle Troubles

Learn about similar triangles by following the directions below.

Angle	Angle measure (cardboard triangle)	angle measure (shadow)
A		
B		
C		

Side	Side length (cardboard)	Side length (shadow)	Cardboard side length/shadow
AB			
BC			
AC			

1. Draw a triangle on a piece of cardboard. Then cut out the triangle and label the vertices with the letters A, B, and C.

2. Measure the sides and the angles of your triangle and record each measurement in the appropriate table above.

3. Tape a piece of string to a vertex of your triangle. Then suspend your triangle 1 foot from a wall. Be sure that the triangle is parallel to the wall.

4. Stand 1 foot back from your triangle and shine a flashlight directly on the triangle to create a shadow of it on the wall. Be sure to hold the flashlight parallel to the floor. (You may need to anchor your triangle with string at the other two vertices to prevent it from moving).

5. Measure each angle and each side of your triangle's shadow. Record each measurement in the appropriate table above.

6. Use a calculator to find the ratios in the last column.

7. What do you notice about the angle measures in the cardboard triangle and in its shadow?

8. What is the relationship between the ratios in the last column of the table?

Comparing triangles

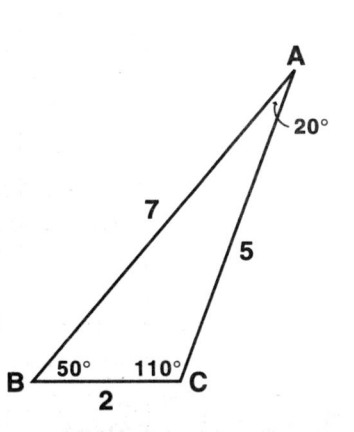

It All Adds Up!

Using a ruler and a protractor, construct the types of triangles in the table below. Label each triangle ABC. Then measure the three sides and three angles of each triangle and fill in the table. Do you notice a pattern in the last three columns? (Hint: See the title of the worksheet.)

Example: scalene
$\overline{AB} = 7$
$\overline{BC} = 2$
$\overline{AC} = 5$
$\angle ABC = 50°$
$\angle ACB = 110°$
$\angle CAB = 20°$

	$\overline{AB}$	$\overline{BC}$	$\overline{AC}$	m∠ABC	m∠BCA	m∠CAB
scalene						
isosceles						
equilateral						
acute						
obtuse						
right						
acute isosceles						
obtuse isosceles						
right isosceles						

Logic, Problem Solving, and Patterning

When the Wright brothers set out to build a flying machine, many intelligent people assured them that their project was a fool's errand. Everybody knew that a machine that was heavier than air could not fly. When Marconi tried to transmit a radio signal from England to Canada, the experts scoffed at the idea of sending radio waves around the curved surface of Earth. When Copernicus and then Galileo proposed that the sun, not Earth, was the center of the known universe, they were condemned as dangerous heretics. Yet each of these people changed the world by defying conventional thinking and by coming up with a radically new and better solution to an existing problem.

Give students guidance by solving several examples of the type of activities in this section before having them complete the activities on pages 56–64.

CONCEPTS

The ideas and activities presented in this section will help students explore the following concepts:
- logic
- problem solving
- patterning

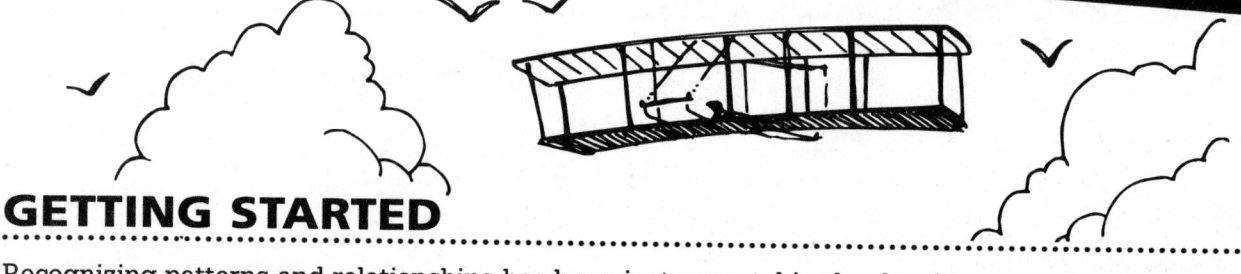

GETTING STARTED

Recognizing patterns and relationships has been instrumental in the development of mathematics and the study of numbers. Exploring patterns and relationships allows students to develop and understand how mathematics applies to their understanding of the development of connections between and among numbers, algebra, geometry, functions, and change relationships. To help students discover patterns and relationships, put the problems below on the board and go over them as a class.

1. Find the pattern.

   3 * 4 → 5   4 * 7 → 1   8 * 4 → 0   1 * 2 → 9

   (Pattern: The sum of the three numbers is 12.)

   Complete the problems below using the pattern above.

   5 * 5 → ___   4 * 1 → ___   6 * 2 → ___

   (2)   (7)   (4)

2. Discover the pattern and find the next two numbers.

   12, 1, 1, 1, 2, 1, 3, 1, . . .

   (Pattern: clock strikes 12 times at noon, 1 time at 12:30, 1 time at 1:00, 1 time at 1:30, 2 times at 2:00, 1 time at 2:30, 3 times at 3:00, 1 time at 3:30; the next two numbers are 4, 1)

3. Discover the pattern and find the next two numbers.

   0, 6, 20, 42, 72, 110, 156, 210, . . .

   (Pattern: 0 x 1 = 0, 2 x 3 = 6, 4 x 5 = 20, 6 x 7 = 42, 8 x 9 = 72, 10 x 11 = 110, 12 x 13 = 156, 14 x 15 = 210; the next two numbers are 16 x 17 = 272 and 18 x 19 = 342)

FUN WITH MATH

Have students explore this number puzzle. Try the puzzle with several 3-digit numbers. Then create a number puzzle to be solved by others using a calculator.

1. Name a 3-digit number (using 3 different digits).
2. Reverse the digits.
3. Subtract the lesser number from the greater number (call this A).
4. Reverse the digits (call this B).
5. Add A to B (call this the result).
    a. What is the result? (1089) Explain why you think this occurs.
    b. Prove that this number will always be the result.

PLAY A GAME

Divide students into groups of 4–6 to play a version of Monopoly™. Give each student a copy of the directions on page 62, a copy of the check register on page 64, and multiple copies of the deposit slip and checks on pages 62 and 63. (Contact local banks to see if they are willing to donate any banking materials.) Before playing, go over how to write checks and fill out deposit slips and how to keep a check register with the students. Also, go over the rules of Monopoly™ in case any student has never played it before.

Chapter Group Project

Divide students into groups. Read the puzzles below to the groups. The students can ask questions about the puzzles. Answer students' questions with either "yes," "no," or "irrelevant." If the group solves the puzzle within 20 minutes, then it scores the rating of 2, 3, or 4. If the group gets stuck, then a clue can be given to the group from the clues section, but a clue costs a point.

Recovery—A truck driver called in to his office to report that his truck had broken down. A tow truck was sent out to tow back the disabled truck. When they arrived, the truck that had broken down was towing the tow truck. Why?
Clues: There was nothing wrong with the tow truck. The truck that had broken down had a serious fault which was remedied by the way they drove back. (Answer: The truck had broken down because its brakes had completely failed. The truck driver drove back towing the recovery vehicle. When he needed to slow, he signaled with his hand and the recovery truck driver applied his brakes, thereby slowing both vehicles.)

Man Overboard—A man on vacation in a different country was alone on his yacht when he fell off into deep water. He was a non-swimmer and was not wearing anything to help keep him afloat. He was rescued half an hour later. Why didn't he drown?
Clue: The previous day, he had bought some beautiful postcards of Jerusalem. (Answer: He fell into the Dead Sea, which lies between Israel and Jordan. The water of the Dead Sea is so salty and dense that anyone in it floats very easily.)

A Riddle—An old riddle goes like this: A man without eyes saw plums on a tree. He did not take plums, and he did not leave plums. How could this be?
Clue: He had sight, and he took fruit. (Answer: The answer lies in the use of plurals. He did not have eyes, he had one eye. He saw two plums on a tree. He took one and left one, so he did not take "plums" or leave "plums.")

LOGIC, PROBLEM SOLVING, AND PATTERNING

FS122009 Pre-Algebra Made Simple • © Frank Schaffer Publications, Inc.

Perplexing Puzzles

Use your logic skills to solve the puzzles below. The charts are provided to help you.

1. Katie, Randy, and Kelly have chosen careers as a teacher, a pediatrician, and a lawyer. If Randy does not like to be around children and Katie cannot stand the sight of blood, who has chosen which career?

	Katie	Randy	Kelly
teacher			
pediatrician			
lawyer			

2. A cat, a small dog, a goat, and a horse are named Angel, Beauty, King, and Rover. Read the clues below to find each animal's name.
   - King is smaller than either the dog or Rover.
   - The horse is younger than Angel.
   - Beauty is the oldest and is a good friend of the dog.

	cat	dog	goat	horse
Angel				
Beauty				
King				
Rover				

3. Four friends stay at the same hotel for a wedding. Each person has a room on a different floor. Katie must ride the elevator down four floors to visit Kurt. Bill is one floor below Wendy. Kurt has a room on the tenth floor. Wendy must ride the elevator up six floors to visit Katie. Who is staying on which floor?

Katie = _____

Bill = _____

Kurt = _____

Wendy = _____

More Perplexing Puzzles

1. Use the chart below to match each child's name with his or her favorite sport. The favorite sports of Brian, Horace, Ian, and Tammy are baseball, hiking, ice skating, and tennis.

   - No person's name begins with the same letter as his or her favorite sport.
   - Tammy and Horace don't like team sports.
   - Horace and Brian don't like cold weather sports.

	baseball	hiking	ice skating	tennis
Brian				
Horace				
Ian				
Tammy				

2. It is believed that the inhabitants of the ancient town of Koenigsberg enjoyed taking walks over the seven bridges that connected the mainland to two islands in the river that ran through the town. Is it possible for the inhabitants of the ancient town of Koenigsberg to start at any point and take a walk that would cross each bridge exactly once, without retracing their steps?

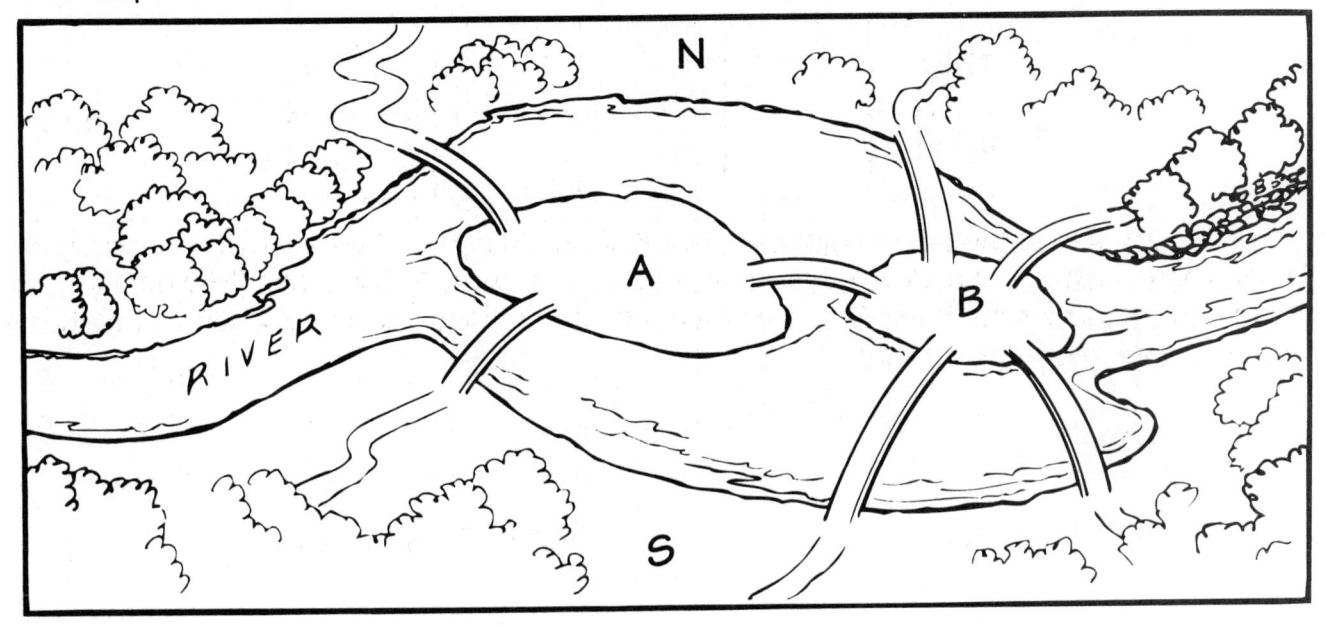

Help the Hungry

Use the chart at the bottom of the page to match the children with their last names, their ages, their streets, and the amount of food they collected.

Two girls (Dawn and Europa) and two boys (Franklin and Gin-Tan) worked all day last Saturday collecting canned food for needy families. The last names of the girls and boys are Qualman, Rockland, Sutton, and Talbot, and their ages are 8, 10, 11, and 12. The streets they live on are Marker Avenue, Nesbitt Road, Oddway Road, and Portland. The number of cans of food collected were 100, 110, 132, and 144.

- Rockland and Gin-Tan's ages are $\frac{1}{10}$ as much as the number of cans of food each collected.
- Gin-Tan collected 100 cans of food.
- Dawn and Sutton are Girl Scouts.
- The boy who lives on Marker Avenue collected more cans than Talbot.
- The number of letters in Europa's street is the same as her age.
- Europa is $\frac{2}{3}$ as old as, and collected more cans of food than, the girl who lives on Oddway Road.

Name_____

Perplexing Problems

Solve the problems below with a partner.

1. The Professors

Two professors of mathematics glared at each other as they examined the same elementary equation. It had been written by a ten-year-old child. "This equation is correct," said one. "No it is absolutely wrong," said the other. How could two experts disagree so completely about a simple equation?

Clues: The two professors each saw a simple, written equation. But, for a very basic reason, they saw it differently. This makes it right for one, but wrong for the other. As they argued about this, they looked straight at each other.

2. Robbery

A gang of criminals was loading a van with television sets that they were stealing from a warehouse when they suddenly heard the siren of an approaching police car. They could not avoid or outpace the police car. How did they escape?

Clues: The gang did not attempt to flee. They thought laterally. The police did not.

3. Murder

An elderly woman is found dead in her bed. She has been murdered. In her bedroom is a fine collection of plates. The police established that she was in good health, seemed perfectly fine when she went shopping the day before, and that no one else had recently visited or entered the house. How did she die?

Clues: The police discovered that she had been poisoned. They checked all the food and drink in the house and could find no trace of the poison. She had bought many fine plates, but rarely went to shops or markets. The day before her death, she had been to the grocery store and the post office.

Fun With Fundraising

The Roadrunners Booster Club planned to sell T-shirts and sweatshirts for their annual fundraiser. They bought 10 T-shirts for the members to wear on the day of the fundraiser at a cost of $75 and 4 sweatshirts for display at a cost of $30. They also spent $85 for their concession stand and $31 for their balloon sales booth. The sales included 19 T-shirts for $760, 2 sweatshirts for $24, snacks totaling $121, and 4 balloons for $16.

Using the above information, solve the problems below.

1. Represent the expenses as signed numbers in a 4 x 1 matrix.

2. Represent the sales as signed numbers in a 4 x 1 matrix.

3. Add the two matrices.

4. Describe the outcome of the four fundraiser categories as profits or losses.

5. Did the fundraiser make an overall profit? What was their overall profit?

A Variety of Patterns

A. Write a variable expression for each pattern. Use n as the variable.

1. 7 • 1, 7 • 2, 7 • 3, 7 • 4, . . .
   _____

2. 4 + 1, 4 + 2, 4 + 3, 4 + 4, . . .
   _____

3. 1 − 1, 2 − 1, 3 − 1, 4 − 1, . . .
   _____

4. 9 − 1, 9 − 2, 9 − 3, 9 − 4, . . .
   _____

5. 60 ÷ 1, 60 ÷ 2, 60 ÷ 3, 60 ÷ 4, . . .
   _____

6. 1 ÷ 5, 2 ÷ 5, 3 ÷ 5, 4 ÷ 5, . . .
   _____

B. When 37 is multiplied by the numbers below, an interesting result happens. Multiply to see this unusual pattern.

37 x 3 = _____

37 x 6 = _____

37 x 9 = _____

37 x 12 = _____

37 x 15 = _____

37 x 18 = _____

37 x 21 = _____

37 x 24 = _____

37 x 27 = _____

C. Solve the problems below to see another unusual pattern.

142,857 x 2 = _____

142,857 x 3 = _____

142,857 x 4 = _____

142,857 x 5 = _____

142,857 x 6 = _____

D. The number 1089 also has many interesting properties. Solve the problems below to find out this interesting pattern.

1089 x 1 = _____

1089 x 2 = _____

1089 x 3 = _____

1089 x 4 = _____

1089 x 5 = _____

1089 x 6 = _____

1089 x 7 = _____

1089 x 8 = _____

1089 x 9 = _____

Patterning 61

Name_____

Play a Game

In this game, you will play Monopoly™ by following the rules on the box. The only difference is that you will use checks for all transactions involving money. Four to six students can play at a time.

1. First, select a banker.
2. Distribute money to players according to the rules of the game. (Give each player a balance of $1500.)
3. Each player must fill out a deposit slip (below) to deposit money in his or her bank.
4. Players keep track of their transactions on a check register (page 64). The banker also keeps track of all the transactions for all the players in the game.
5. Players may keep some small bills to pay fines. Otherwise, all transactions must be completed by check (page 63).
6. Players reconcile their bank statements with their records.
7. The winner is determined according to the rules of the game. The assessment of this activity is based on your banking records and whether the checks and deposit slips were filled out correctly.

_____

_____

DATE _____

_____
Sign here for cash received (if required).

Anytown Bank
Anytown, U.S.A.

Account Number_____

CASH	CURRENCY		
	COIN		
CHECKS			
TOTAL			
LESS CASH RECEIVED			
NET DEPOSIT			

Problem solving

Name_____

Play a Game continued

_____ Check #_____

_____ Date _____

Pay to the order of _____ $_____

_____ Dollars

Anytown Bank
Anytown, U.S.A.

Memo _____ Signature _____

Account Number_____

_____ Check #_____

_____ Date _____

Pay to the order of _____ $_____

_____ Dollars

Anytown Bank
Anytown, U.S.A.

Memo _____ Signature _____

Account Number_____

_____ Check #_____

_____ Date _____

Pay to the order of _____ $_____

_____ Dollars

Anytown Bank
Anytown, U.S.A.

Memo _____ Signature _____

Account Number_____

FS122009 Pre-Algebra Made Simple ■ © Frank Schaffer Publications, Inc.

Play a Game continued

Date	Transaction	Check #	Description	Deposit	Withdrawal	Balance

Problem solving

Statistics and Probability

All students will greatly benefit from the activities involving probability and statistics in this section. Probability and statistics are very visible in many real-life applications. Students are often excited by statistics. In addition to doing the various activities by hand, statistics lends itself to the use of technology.

Try to link everyday situations to your students in which they can use their new skills. For example, students can calculate the probability of winning a race, average their grades, or account for many different combinations of outcomes. Then let them complete the student activity pages (pages 67–74).

CONCEPTS

The ideas and activities presented in this section will help students explore the following concepts:

- mean
- mode
- bar graph
- scatter plot
- probability
- line of best fit
- median
- range
- box and whisker graphs
- histogram
- percent chance
- counting outcomes

GETTING STARTED

Tell students the situation below. Then ask students how they would solve the problem.

NCB stock sold for 14 ½ dollars per share on Monday morning, but by noon, its value had risen 2 ¼ dollars per share. By closing time the same day, the stock had dropped ½ of a dollar per share. What was the closing value of the stock? ($16.25)

THE STOCK MARKET

As a class, choose one stock to follow for a three-week period. Tell the students the initial investment and have them write it in chart form in a journal. (See example below.) Each student is responsible to check the stock price daily and record it on the chart. Students should calculate the sellout price daily and determine if the class would profit or lose by selling that day.

Stock	Buy in price	# of shares purchased			
MEMC	5.95	420.17			

Date	Stock price	# of shares	Sellout price	Initial investment	Profit/loss
9/6/98	6.05	420.17	2542.03	2500	+42.03
9/7/98	5.98	420.17	2512.62	2500	+12.62

Follow-Up Activity: Have groups of four students choose a stock to purchase. Students will follow the stock for two weeks, keeping track of daily profits/losses. Have students write a one- to three-paragraph essay discussing the movement of the stock over the two-week time frame. The paragraphs should include specifics about profits/losses and a graphic representation of this information. Students could also write a three- to five-paragraph essay about the stock they choose. It could include when the company joined the stock market and with which exchange the company is listed. Encourage students to discuss the number of exchanges that exist and characteristics of them.

Internet Link: As a class, take part in the on-line stock market simulation on the Web page "Taking Stock" at http://www.santacruz.k12.ca.us/~jpost/projects/TS/TS.html. Students will not only gain an understanding of the stock market with this activity, but they will also learn about fractions of a dollar, stock symbols, profit and loss, corporate research, and persuasive letters.

FUN WITH STATISTICS

One of America's favorite pastimes is playing board games. Let students work together in groups to create a board game of their favorite sport. The rules can be made up however each group decides. The idea is to have students create rules based on statistical probabilities they have observed through research or observation. For example, a baseball game might include the following rules for getting a hit: *If the batter answers a math question correctly, the player will roll 1 (or 2) die to determine if a hit, out, or walk occurs. A 1, 2, or 3 will result in a hit; a 4 or 5 will result in an out; a 6 will result in a walk. If a hit occurs, then a second roll will determine how many bases the batter will receive.* As you can see, these games could range from simple to very complex. These games, once created, can be useful for reviewing for a test, creating tournaments for practice, or just plain fun! Let each group take turns playing each game.

WRITING FUN

Have students list five occupations that they think never use mathematics and/or computers. As a class, list the mathematics actually used in an occupation that seems mathematics-free, such as a gift-wrapper in a department store. Next, have students write a paragraph describing the career they might want to pursue and the use of mathematics and/or computers involved in that occupation.

Chapter Group Project

Have students complete Parts 1–5 below to learn more about statistics and the Boston Marathon.

Part One: Divide students into groups of four. Have students use the Internet to gather information about the Boston Marathon (two web sites are http://www.bostonmarathon.org and http://www.baa.org) and write one to three paragraphs about the history of the race.

Part Two: Have groups find the men's winning times. Students then convert the winning times from hours:minutes:seconds into minutes. Using the information, students create an XY scatterplot.
Example:

Year	Winning time hours:minutes:seconds	Convert hours to minutes	Convert seconds to minutes	Total time in minutes
1897	2:55:10	120	0.17	175.17
1898	2:42:00	120	0.00	162.00
1899	2:54:38	120	0.63	174.63
1900	2:39:44	120	0.73	159.73

Part Three: As a class using Excel or the TI-83, draw the line of best fit onto the scatterplot and determine the equation of the line. (Assume the data is linear.)

Part Four: Have groups predict the winning time of the next race using the equation of the line of best fit. Next, have the groups predict the winning times for 2000, 2005, and 2050.

Part Five: Have the groups discuss their predictions and write conjectures as to the soundness of their predicted winning times.

Name _____

The Stock Market

Use the stock market section of the newspaper to select two stocks of companies of your choice to follow. Using the charts below, keep track of the stocks for one week. At the end of the week, analyze the data to determine the trends for increases and decreases of your two stocks.

Name of stock _____

Buy in price _____ # of shares purchased _____ initial investment $_____

Date	Stock price	# of shares	Sellout price	Profit/loss

Name of stock _____

Buy in price _____ # of shares purchased _____ initial investment $_____

Date	Stock price	# of shares	Sellout price	Profit/loss

Statistics

Name _____

A Mean Way to Estimate

Find the mean, median, mode, and range for the sets of data below.

1. 1, 2, 3, 4, 5, 6, 6, 5, 6, 5, 4

    mean _____

    median _____

    mode _____

    range _____

2. 10, 10, 11, 16, 15, 10

    mean _____

    median _____

    mode _____

    range _____

3. 100, 125, 135, 115, 200

    mean _____

    median _____

    mode _____

    range _____

4. 0.1, 0.3, 0.5, 0.9, 0.3, 0.6, 0.3, 0.9

    mean _____

    median _____

    mode _____

    range _____

5. 1000, 2000, 3000, 1000, 1000

    mean _____

    median _____

    mode _____

    range _____

6. 23, 34, 56, 78, 90, 23, 34, 12, 12, 12

    mean _____

    median _____

    mode _____

    range _____

7. 45, 75, 15, 0, 35, 60, 30

    mean _____

    median _____

    mode _____

    range _____

8. 10, 18, 7, 3, 31, 5, 13, 10, 6

    mean _____

    median _____

    mode _____

    range _____

Mean, median, mode, and range

FS122009 Pre-Algebra Made Simple • © Frank Schaffer Publications, Inc.

Soccer Statistics

Do this activity using the TI-83 graphing calculator.

The Nashville Metros Soccer team has 18 players. Their ages are 32, 25, 22, 23, 32, 27, 23, 20, 21, 27, 22, 22, 23, 29, 22, 20, 22, and 27.

> Follow the directions below using your TI-83. Then answer questions 1–7.
>
> [STAT] [EDIT] 1:Edit [ENTER]
>
> Type data into List one.
>
> [2nd] [mode] [STAT] arrow over to Calc [ENTER] 1:1 — var stats [ENTER] [ENTER]
>
> You will find the mean, range, quartiles.
>
> To make a box and whisker graph, change window settings by:
>
> [WINDOW] x-min = 0, x-max = 50, x-slc = 1, Y-min = 0, Max = 30, Y-scl = 1
>
> [2nd] [Y=] 1: plot 1 [ENTER] Turn stat plot on choose whisker box [GRAPH]
>
> To sort the data to find the mode and median [STAT] 2 sort A (2nd 1 [ENTER]
>
> [STAT] edit 1:edit [ENTER] . Use the sorted data to find the median and mode.

1. Find the median age. Is it one of the ages?

2. Determine the mean of the ages.

3. What is the mode?

4. What is the range of the ages?

5. Determine the upper and lower quartiles.

6. Determine the upper and lower extremes.

7. On a sheet of graph paper, make a box and whisker graph using this information.

Name_____

Graphs of All Kinds

Using the sets of data in A–D below, create four graphs. You must have at least one line graph, one line plot, one bar graph, and one histogram.

A. Compare the temperature in Honolulu and Miami in August.

	Honolulu	Miami
August 12	89	89
August 14	90	91
August 16	91	88
August 18	89	91
August 20	89	90

B. There are 20 students in a geometry class. Their ages are as follows:

14, 15, 16, 17, 18, 16, 16, 14, 14, 15, 15, 16, 14, 16, 15, 17, 15, 18, 15, 15

After you create a graph, find the mean, median, and mode of their ages.

mean _____ median _____ mode _____

C. Below is the median household income for the states listed.

	1987	1993
California	$37,231.00	$33,083.00
Connecticut	$40,586.00	$38,369.00
Indiana	$27,812.00	$28,618.00
Texas	$30,531.00	$27,892.00
Utah	$32,764.00	$34,746.00

D. Rafaela owns two shoe stores. The profits from *Rafaela's I* during the years 1991–1995 were $12,000; $16,000; $16,500; $18,000; and $20,500. The profits from *Rafaela's II* over the same period were $16,000; $14,500; $17,000; $16,000; and $14,000. Which store is more profitable and how is this revealed by the graph?

Mean, median, mode, and graphing

Name_____

How Hot Is It?

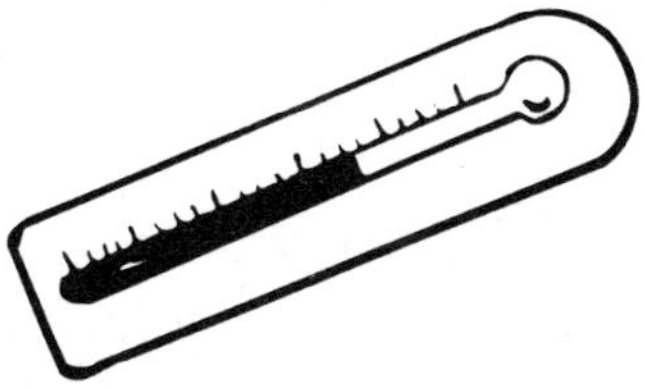

Using the information in the table below, follow the directions.

Average Daily High Temperatures

	Jan.	Feb.	Mar.	April	May	June	July	Aug.	Sept.	Oct.	Nov.	Dec.
Fort Worth	56	61	67	75	84	90	97	97	89	80	69	59
Houston	63	66	72	78	85	92	93	93	90	84	73	67
San Antonio	62	65	72	79	85	92	96	96	90	81	70	64

1. Graph the average daily high temperatures of the three cities. Use a different colored pencil for each city.

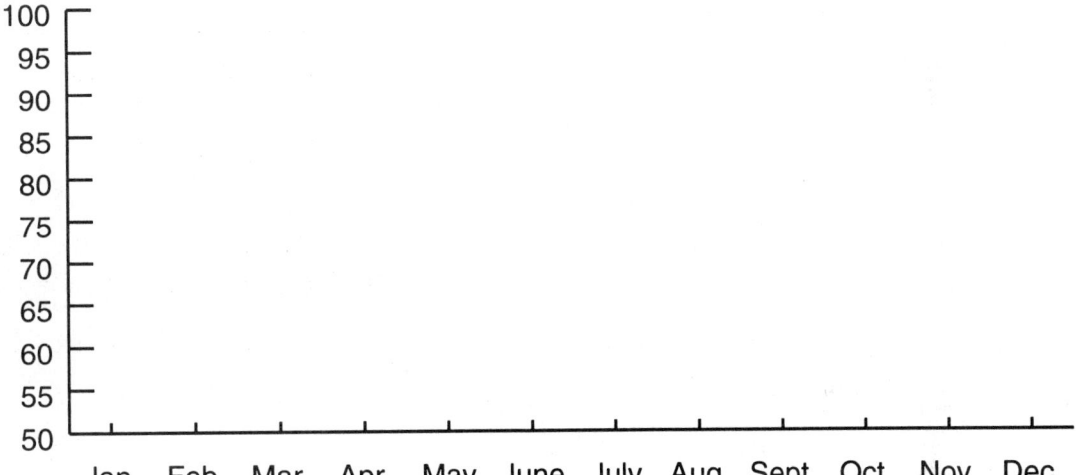

2. What is the average temperature for each city?

   Fort Worth = _____

   Houston = _____

   San Antonio = _____

3. What is the range of temperatures for each city?

   Fort Worth = _____

   Houston = _____

   San Antonio = _____

4. What is the median temperature for each city?

   Fort Worth = _____

   Houston = _____

   San Antonio = _____

Line graphs and inequalities

Name_____

Rolling a Die

With a partner, follow the directions below to learn all about probability.

1. Roll a die 50 times and record the results (in tally form) on the chart at the bottom of the page. Then count the tally marks and write the totals in the third column.

2. Determine the possibility of rolling each number. To do this, divide the total number column of each die number by 50. Write the answers in the chart. Is it all about the same? _____ Explain your answer.

3. Find the sum of all the probabilities and write it in the chart. Explain why it is close to (or equal to) 1.

4. Find the percent chance of each event. Find the sum of the percent chance. Explain this sum.

Die #	# of Tally Marks	Total #	Probability	% Chance
1				
2				
3				
4				
5				
6				
Totals		50		

Name_____

Delicious Decisions

Follow the directions below to learn about predicting outcomes.

1. Label five index cards with the food and juice items on the menu to the right.

   > **juices:** orange, grapefruit
   > **main dishes:** poached eggs, cold cereal, hot cereal

2. Suppose that you are going to have a meal consisting of one item from each category on the menu in Step 1.

   a. How many juices do you have to choose from? _____

   b. How many main dishes do you have to choose from? _____

   c. Use your index cards to help you make a list of all the different meals you could choose. How many choices are there altogether? _____

3. What is the relationship between your answers to parts a–c of Step 2?

4. Label two more index cards with the side dishes on the menu to the right.

   > **juices:** orange, grapefruit
   > **main dishes:** poached eggs, cold cereal, hot cereal
   > **side dishes:** whole wheat toast, bran muffin

5. Suppose that you are going to have a meal consisting of one item from each category on the menu in Step 4.

   a. How many juices do you have to choose from? _____

   b. How many main dishes do you have to choose from? _____

   c. How many side dishes do you have to choose from? _____

   d. Use your index cards to help you make a list of all the different meals you could choose. How many choices are there altogether? _____

6. What is the relationship between your answers to parts a–d of Step 5?

7. In general, what do you think is the relationship between the number of choices in each category and the number of different combinations that include one item from each category?

FS122009 Pre-Algebra Made Simple ▪ © Frank Schaffer Publications, Inc.

Counting outcomes 73

Name_____

Super Scores

Data that can be organized in a frequency table with intervals can also be displayed in a histogram. A histogram resembles a bar graph, but a histogram has no spaces between consecutive bars. Construct a histogram by following the directions below.

1. Cut out the boxes of the winning football scores at the bottom of the page.
2. Place one box above the appropriate interval on the graph below for each winning score. Do not leave any space between columns.
3. Using a pencil, trace around each column of boxes that you made in Step 2. Then remove the boxes to reveal your histogram.
4. Using the histogram, answer the questions below.

   a. How many winning scores are displayed in the histogram? _____
   b. Which interval contains more winning scores, 11–20 or 31–40? _____
   c. In how many football games was the winning score 11–20 points? _____
   d. In how many football games was the winning score greater than 40 points? _____
   e. In how many football games was the winning score less than 31 points? _____

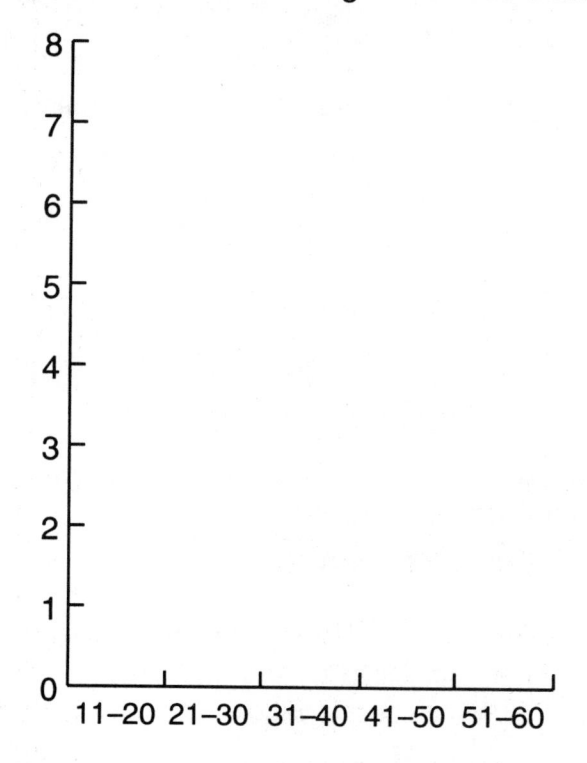

(Histogram with y-axis labeled 0–8 and x-axis intervals: 11–20, 21–30, 31–40, 41–50, 51–60)

| 35 | 32 | 39 | 14 | 27 | 16 | 35 | 20 | 16 | 38 | 16 | 27 | 20 |
| 24 | 26 | 33 | 27 | 42 | 24 | 38 | 23 | 31 | 55 | 21 | 46 |

Constructing a histogram

Answer Key

Page 3
K. 319	G. 7	S. -98
D. -29	B. -39	C. 67
I. 62	N. 65	Y. -54
V. -1454	F. 18	U. 384
A. 17	P. -260	L. -19
T. -179	M. -129	W. -70
E. 44	R. 312	X. -372
Q. -91	O. 97	H. -325

EUREKA; DIRIGO (I direct); EXCELSIOR; HOPE; FRIENDSHIP; INDUSTRY; FORWARD

Page 4
1. 60
2. 325
3. 1995
4. 2940
5. Answers will vary.
6. Ben
7. Jan
8. Jena
9. Andy

Page 5
1. $48
2. $92
3. $34
4. $18
5. $60
6. $63
7. $12
8. $327
9. $347
10. $10
11. $60
12. $70
13. $80

Page 6
1. $189; $160; $8320
2. $456; $369; $19,188
3. $320; $264; $13,728
4. $570; $432; $22,464
5. $228; $192; $9984
6. $296; $255; $13,260
7. $12,948
8. $5356
9. $12,532
10. $280

Page 7
1. $2136; $178
2. $1632; 136
3. $3528; $147
4. $5868; $163
5. $3756; $313
6. $2136; $89
7. $4944; $206
8. $247.92
9. $225
10. $146.25
11. 230.88

Page 8
1. $64
2. $55.50
3. $128.80
4. $97.65
5. 864 mi
6. 864 mi
7. 26 m.p.g.
8. 17 m.p.g.

Page 9
1. 3 • (8 − 3) = 15
2. (4 • 5) + 4 = 24
3. 4 • (9 − 6) = 12
4. (3 • 7) + 0 = 21
5. 4 + (6 • 2) = 16
6. 3 • (9 − 1) = 24

Page 10
1. 5
2. 4
3. 6
4. 8
5. 9
6. 20
7. 12
8. 3
9. 63
10. 11

Page 11
1. 2 • 4 − 6
2. 2 • 6 − 4
3. 4 • 6 − 2
4. (2 + 4) ÷ 6
5. (6 + 4) ÷ 2
6. (6 + 2) ÷ 4
7. (6 + 4) • 2
8. (2 + 6) • 4
9. (2 + 4) • 6
10. 6 ÷ (4 + 2)

Page 12
1. 2 • 3 ÷ 6
2. 2 • 4 − 3
3. 4 ÷ 2 • 3
4. 4 • 6 ÷ 3
5. 6 • 3 − 4
6. 6 • 2 + 3

Page 13
1. 15 ÷ 5 − 3
2. 15 − 12 + 5
3. 5 • 3 ÷ 15
4. 12 ÷ 3 • 5
5. 12 − 15 ÷ 3
6. 15 − 5 + 3

Page 14
1. $11/12 − 5/12$
2. $1/12 + 11/12$
3. $7/12 − 5/12$
4. $7/12 + 11/12$
5. $5/12 − 1/12$
6. $7/12 + 1/12$

Challenge $11/12 + 1/12 − 7/12 − 5/12$

Page 15

10	−	4	−	4	=	2
−		+		+		
2	+	8	+	1	=	11
−		−		−		
7	+	2	+	3	=	12
=		=		=		
1		10		2		

7	+	4	−	6	=	5
−		+		+		
6	+	7	−	5	=	8
−		−		−		
5	+	3	−	8	=	0
=		=		=		
8		8		3		

Page 16

9	x	1	+	7	=	16
x		x		x		
5	x	6	+	8	=	38
÷		÷		÷		
3	x	2	+	4	=	10
=		=		=		
15		3		14		

5	x	4	÷	2	=	10
x		+		+		
6	x	9	÷	3	=	18
÷		−		+		
5	+	7	−	8	=	4
=		=		=		
6		6		13		

Page 17
Across: 45, 45, 63, 49, 55, 49, 58, 52, 52, 53, 521
Down: 59, 60, 53, 69, 49, 47, 56, 66, 62, 521
Just for Fun! Answers will vary.

Page 18
1. $\frac{1}{80} = \frac{1¾}{x}$; x = 140 miles
2. $\frac{1.08}{12} = \frac{x}{8}$; x = $.72
3. $\frac{1}{4} = \frac{x}{60}$; x = 15
4. $\frac{12}{4.99} = \frac{1}{x}$; x = $.42

Page 19
1. 20%
2. 90%
3. 45
4. 8.82
5. 75
6. 300
7. 11%
8. 130,000

Page 20
1. $\frac{44}{100} = \frac{x}{17.50}$; x = $7.70
2. $\frac{41}{100} = \frac{x}{420}$; x = $172.20
3. $\frac{x}{100} = \frac{1495}{1649}$; x = 90.7%
4. $\frac{x}{100} = \frac{57}{119}$; x = 47.9%
5. $\frac{81}{100} = \frac{19.96}{x}$; x = $24.64
6. $\frac{67}{100} = \frac{4.97}{x}$; x = $7.42

Page 21
1. 9:15 a.m.
2. 6:15 p.m.
3. 12:25 p.m.
4. 9:45 a.m.
5. 9:15 a.m.

Time	Hours	Hours to Minutes	Minutes	Seconds	Seconds to Minutes	Total Time in Minutes
Ex. 2:17:34	2	2•60	120	17	34÷60	.57	137.57
6. 3:31:22	2•60	180	31	22÷60	.37	211.37
7. 4:16:12	3•60	240	16	12÷60	.20	256.20
8. 2:26:31	4•60	120	26	31÷60	.52	146.52
9. 2:34:20	2•60	120	34	20÷60	.33	154.33
10. 2:41:72	2•60	120	41	72÷60	1.20	162.20
11. 2:16:21	2•60	120	16	21÷60	.35	136.35
12. 2:10:43	2•60	120	10	43÷60	.72	130.72
13. 2:08:60	2•60	120	8	60÷60	1	129
14. 2:04:43	2•60	120	4	43÷60	.72	124.72
15. 2:04:08	2•60	120	4	8÷60	.13	124.13

Page 22
A. 1. $286.44; $252.50; $234.11; $205.11; $185.13; $148.84
    2. Graphs will vary.
B. 1. $17.85 ÷ 3 = $5.95 each or $6.50 each. Buying 3 is a better deal.
    2. $.55 savings per shirt or $1.65 total
    3. $35.70 or 101,745 pesos
    4. $2.30 or 6555 pesos

Page 23
Answers will vary.

Page 24
A. 1. 400 cards
    2. at least 401
    3. Answers will vary.
B. 1. 8:30 a.m., $15.00 each
    2. heavy equipment operator = $105; laborer = $70
    3. heavy equipment operator = $131.25; laborer = $95

Page 25
1.

# of tickets	Discount per ticket	Adult unit cost per ticket	Youth unit cost per ticket
10 to 20	$1.50	$6.50	$4.50
21 to 40	$1.75	$6.25	$4.25
41 to 60	$2.00	$6.00	$4.00
61 to 80	$3.50	$4.50	$2.50
81 to 100	$3.75	$4.25	$2.25
100+	$4.00	$4.00	$2.00

2. $75.97
3. $6.00; $350.63
4. $6.25; $71.72; $337.35
5. $2.13
6. a. $\frac{1.50}{8.00}$ b. $\frac{1.75}{8.00}$ c. $\frac{2.00}{8.00}$
   d. $\frac{3.50}{8.00}$ e. $\frac{3.75}{8.00}$ f. $\frac{4.00}{8.00}$
7. a. 25% b. 29% c. 33%
   d. 58% e. 62.5% f. 67%
8. 19–50%; 25–67%

Page 26
1. $55.25
2.–3. Answers will vary.

Page 27
1. 5 + 3 • 6
2. 5 • 7 + 2 + 2
3. 8 − 4 + 1
4. $\frac{24 + 16}{15 - 10}$

5. $3^2 + [(10 ÷ 5) − 2]$
6. (17 − 2) • 10 − 8
7. $(8^2 + 12)$ • 3 ÷ 3 • 4
8. [(9 + 5) • 2 − 16] − 5
9. $[75 − 21 ÷ (5 − 2) • 3^2]$ • 5
10. 4^2 • 7 + (12 − 3)

Page 30
Activity I
1.–4. Answers will vary.
Activity II
1. Students should place 6 green and 12 red M&M's™ on their desks.
2. a. 2(3n + 6)
   b. 3(2n + 4)
   c. 6(n + 2)
3. They are equivalent expressions.
4. 2(9n + 12), 3(6n + 8), 6(3n + 4)

Page 31
1. a. 3g + 5o + 6b
   b. g + 5o + 3b
2. a. 5b + 7r + 3b + 8g
   b. yes; 8b + 7r + 8g
   c. 2b + 4r
3. 5g + 5r
4.–6. Answers will vary.
7. a. 7y + 5o + 6b
   b. 5y + 5o − 6b (borrowed 6 brown from the bag)

Page 32
1. a^9
2. m^{11}
3. x^{14}
4. b^{14}
5. y^{17}
6. $30n^5$
7. $21x^3y^2$
8. $28d^{15}$
9. $30x^8$
10. $36m^8$
11. $27a^4x^3$
12. a^{17}
13. $56a^{12}$
14. $42b^9$
15. $45m^{15}$
16. $27c^9$
17. $30a^{15}$
18. $84a^{11}$
19. $126c^3d$
20. $280x^2y$
21. $20c^2d$
22. $270x^{18}$
23. $21c^8d^2$
24. $160m^9x^3$
25. $240a^9$
26. $480c^4d^{10}$
27. $90w^5y^3$
28. a^8b^8
29. $m^{13}n^{15}$
30. x^{22}
31. $120a^3x^6$
32. $240a^6d^{10}$
33. $180x^{12}y^6$

Page 33
Answers will vary. Possible answers are below.
1. 10 + 10 + 4, 6 + 6 + 12, 5 + 5 + 14
2. 5 + 10 + 12, 4 + 11 + 12, 3 + 14 + 10
3. 10 + 6, 12 + 4, 13 + 3
4. 10 + 10 + 6 + 6, 9 + 9 + 7 + 7, 12 + 12 + 4 + 4
5. 5 + 5 + 5 + 6, 6 + 6 + 6 + 3, 4 + 4 + 4 + 9
6. 6 + 6 + 6 + 6 + 1, 4 + 4 + 4 + 4 + 9, 3 + 3 + 3 + 3 + 13
7. 5 + 5 + 2, 3 + 3 + 6, 2 + 2 + 8
8. 8 + 8 + 8 + 3 + 3, 4 + 4 + 4 + 9 + 9, 2 + 2 + 2 + 12 + 12

Page 34
A. 9   E. 540   T. 0
R. ⁻45   I. ⁻3   G. 402
C. ⁻1000   H. ⁻31   N. 14
I. 60   T. 42   T. ⁻210
S. 4   D. ⁻65
THE TITANIC

Page 35
1. x = 5
2. m = ⁻12
3. x = 1
4. t = 6
5. n = 2
6. a = 24
7. r = ⁻229
8. m = 17
9. k = 12
10. y = 4

Page 36
1. LCM = 2; GCF = 4
2. LCM = 2; GCF = 20
3. LCM = 2; GCF = 40
4. LCM = 2a; GCF = 18a
5. LCM = 2x; GCF = 4x
6. LCM = 5m; GCF = $5m^2$
7. LCM = 3; GCF = 9
8. LCM = 3; GCF = 3
9. LCM = 2; GCF = 2
10. LCM = m; GCF = m

Page 39
Answers will vary.

Pages 40–41
1. circle
2. hexagons and pentagons
3. yes
4. hexagon = 9, pentagon = 5

5. hexagon = 720, pentagon = 540
6. 71 cm
7. d = 8.9 in, d = 22.6 cm
8. r = 4.45 in, r = 11.3 cm
9. SA = 55.89 in, SA = 141.93 cm
10. V = 18.63 in, V = 47.31 cm
11. 91–118.3 m; 45.5–91 m
12. 300 < P < 460
13. 5000 < A < 13,000
14. 8 yd, 7.28 m; 2.43 m
15. 120 yd^2
16. 792 yd^2
17. C = 62.8 yd; A = 314 yd
18. Answers will vary.

Page 46
2. yes; yes
3. yes; 3 (one from each vertex)
5. 4
6. pentagon—5; hexagon—6
7. number of sides of any regular polygon equals the number of lines of symmetry

Page 47
5. They are all the same.
6. 3.141
7. 3.14; C ≈ πd

Page 48
Answers will vary.

Page 49
6. V = πr^2h (Bh)
7. V = ⅓πr^2h (⅓Bh)

Page 50
1.

Face	Area
A	lw
B	lh
C	wh
D	lh
E	wh
F	lw

2. rectangular prism (rectangular box)
3. B = D; A = F; C = E
4. A, B, C
5. 2A + 2B + 2C (2lw + 2lh + 2wh)
6. l = 8.5 cm; w = 4.8 cm; h = 3.5 cm
7. 174.7 cm^2

Page 52
2. Answers will vary.
5. Answers will vary.
6. Answers will vary.
7. Similar triangles have congruent corresponding angles.
8. The lengths of the corresponding sides of similar triangles are in proportion.

Page 53
Triangles will vary. The sum of the interior angles of all triangles is equal to 180°.

Page 56
1. Katie—teacher; Randy—lawyer; Kelly—pediatrician
2. cat—King; dog—Angel; horse—Rover; goat—Beauty
3. Katie—14th; Kurt—10th; Bill—7th; Wendy—8th

Page 57
1. Brian hikes; Horace plays tennis; Ian plays baseball; Tammy ice skates.
2. No, it is not possible.

Page 58
Dawn Qualman, 12 years old, 132 cans, Oddway Road
Europa Sutton, 8 years old, 144 cans, Portland
Franklin Rockland, 11 years old, 110 cans, Marker Avenue
Gin-Tan Talbot, 10 years old, 100 cans, Nesbitt Road

Page 59
1. The equation was 9 x 9 = 81, but they were looking at it from opposite sides of the table. So to one professor it was correct, but to the other, it read 18 = 6 x 6, and so it was incorrect.
2. They started to unload the television sets and carried them back into the warehouse. When the police arrived, the robbers told the police that they were making a late delivery and they were believed!
3. The elderly woman was poisoned by her greedy nephew, who wanted to inherit her fortune. He sent her what looked like a mailer with a fantastic offer for a collector plate which he knew she would want to have. To order the plate, the offer had to be completed, folded, and sealed and sent back without delay. The nephew had put a slow-acting poison on the seal of the mailer. Once his aunt had licked the seal and posted the mailer, there was nothing to connect him with the crime.

Page 60
1. $\begin{bmatrix} -75 & -85 \\ -30 & -31 \end{bmatrix}$ 2. $\begin{bmatrix} 760 & 121 \\ 24 & 16 \end{bmatrix}$ 3. $\begin{bmatrix} 685 & 36 \\ -6 & -15 \end{bmatrix}$

4. $685 profit on T-shirts, $36 profit on concession sales, $6 loss on sweatshirts, and $15 loss on balloon sales
5. overall profit of $700.00

Page 61
A. 1. 7 • n   2. 4 + n
   3. n – 1   4. 9 – n
   5. 60 ÷ n   6. n ÷ 5
B. 111, 222, 333, 444, 555, 666, 777, 888, 999
C. 285,714; 428,571; 571,428, 714,285; 857,142 (All have the same digits in the same order as the original number with a different starting point.)
D. 1089, 2178, 3267, 4356, 5445, 6534, 7623, 8712, 9801

Page 67
Answers will vary.

Page 68
1. mean = 4.3, median = 5, mode = 5, 6, range = 1–6
2. mean = 12, median = 10.5, mode = 10, range = 10–11
3. mean = 135, median = 125, mode = none, range = 100–200
4. mean = 0.4875, median = 0.4, mode = 0.3, range = 0.1–0.9
5. mean = 1600, median = 1000, mode = 1000, range = 1000–3000
6. mean = 37.4, median = 28.5, mode = 12, range = 12–90
7. mean = 37.14, median = 35, mode = none, range = 0–75
8. mean = 11.4, median = 10, mode = 10, range = 3–31

ANSWER KEY

Page 69
1. 23, yes
2. 24.4
3. 22
4. 20–32
5. 22, 27
6. min = 20, max = 32
7. Check students' graphs.

Page 70
A. Graphs will vary.
B. Graphs will vary; mean = 15.5; median = 15; mode = 15
C. Graphs will vary.
D. *Rafaela's I* is more profitable.

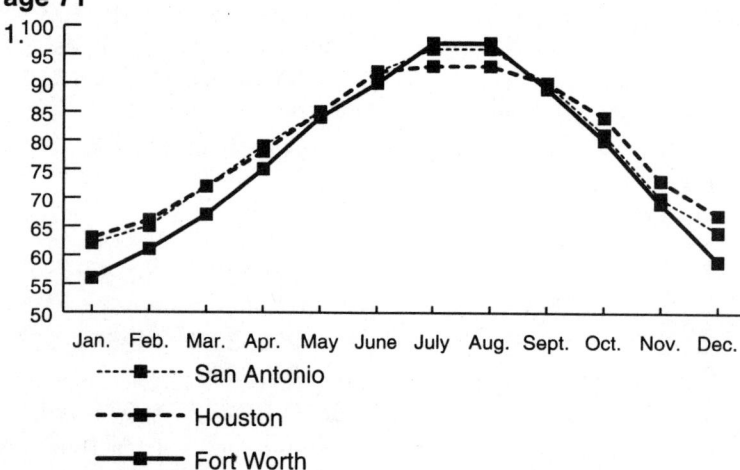

Page 71

2. Fort Worth = 77°
   Houston = 79.7°
   San Antonio = 79.3°
3. Fort Worth = 56°–97°
   Houston = 63°–93°
   San Antonio = 62°–96°
4. Fort Worth = 77.5°
   Houston = 81°
   San Antonio = 80°

Page 72
Answers will vary.

Page 73
2. a. 2; b. 3; c. 6
3. juices x main dishes = number of choices (2 x 3 = 6)
5. a. 2; b. 3; c. 2; d. 12
6. juices x main dishes x side dishes = number of choices (2 x 3 x 2 = 12)
7. Multiply all possibilities together.

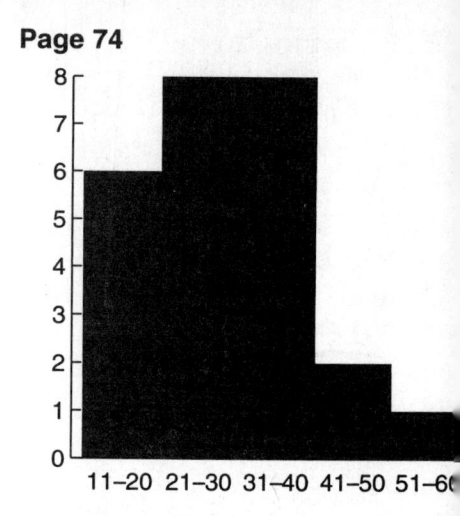

Page 74

4. a. 25
   b. 31–40
   c. 6
   d. 3
   e. 14

FS122009 Pre-Algebra Made Simple ▪ © Frank Schaffer Publications, Inc.